Wireless Networks

Series Editor

Xuemin Sherman Shen, University of Waterloo, Waterloo, Canada

The purpose of Springer's Wireless Networks book series is to establish the state of the art and set the course for future research and development in wireless communication networks. The scope of this series includes not only all aspects of wireless networks (including cellular networks, WiFi, sensor networks, and vehicular networks), but related areas such as cloud computing and big data. The series serves as a central source of references for wireless networks research and development. It aims to publish thorough and cohesive overviews on specific topics in wireless networks, as well as works that are larger in scope than survey articles and that contain more detailed background information. The series also provides coverage of advanced and timely topics worthy of monographs, contributed volumes, textbooks and handbooks.

Xin Wei · Dan Wu · Liang Zhou

Cross-Modal Communication Technology

Springer

Xin Wei
Nanjing, China

Dan Wu
Nanjing, China

Liang Zhou
Nanjing, China

ISSN 2366-1186 ISSN 2366-1445 (electronic)
Wireless Networks
ISBN 978-3-032-00956-2 ISBN 978-3-032-00957-9 (eBook)
https://doi.org/10.1007/978-3-032-00957-9

This Springer imprint is published by the registered company Springer Nature Switzerland AG
The registered company address is: Gewerbestrasse 11, 6330 Cham, Switzerland

Preface

With the rapid development of multimedia communications, humans try to pursue multi-dimensional and higher-level sensory experiences after their audio-visual demands have been greatly satisfied. Currently, haptic information is gradually being integrated into traditional audio-visual-dominant multimedia services, forming multi-modal services. Multi-modal services have been considered killer applications in the B5G and 6G eras, including remote industrial robotic grasping, tele-surgery and diagnosis, online immersive shopping and gaming, etc. Correspondingly, the supporting communication technologies have evolved from traditional multimedia communications and the recently much-discussed tactile internet to cross-modal communications. Compared with traditional multimedia communication and the tactile internet, the cross-modal communication paradigm is characterized by the collaborative transmission and processing of audio, visual, and haptic signals and streams. By fully exploring the potential correlations among audio, visual, and haptic modalities, it enhances natural human-machine interactions and immersive experiences in multi-modal services while meeting the requirements of low latency, high reliability, and high throughput.

Since 2020, the MultiMedia Communication (MMC) Research Group founded by the authors has conducted a series of systematic and in-depth studies on the concept, architecture, key techniques, and applications of cross-modal communications, achieving significant academic influence. This book is a summary and compilation of the research works of the MMC Research Group in this field over the past five years. It is also the first book on cross-modal communications. We hope that through the descriptions of this book, readers will gain a deep understanding of the development and technical details of this new communication paradigm for the emerging multi-modal services.

This book is authored by Prof. Xin Wei and Prof. Liang Zhou from Nanjing University of Posts and Telecommunications, and Prof. Dan Wu from Army Engineering University of PLA. It can serve as a good reference for scientific researchers and technical personnel in the fields of multimedia communications, multimedia signal processing, artificial intelligence, etc. It can also be used as a textbook or supplementary reading material for senior undergraduate and graduate students majoring in electronic engineering, computer science and technology, pattern recognition and artificial intelligence, etc.

The authors have always emphasized the close integration of theory and application during the writing process and have endeavored to introduce all aspects of cross-modal communication technology to readers in a concise, easy-to-understand, and accessible way. However, it is inevitable that there may be shortcomings due to our limited expertise. The authors sincerely welcome criticism and correction from readers.

Nanjing, China
November 2025

Xin Wei
Dan Wu
Liang Zhou

Acknowledgements

This work is partly supported by the National Natural Science Foundation of China (Grants Nos. 62427802, 62231017, 62277032, 62071254), the Higher Education Reform Foundation of Jiangsu Province (Grant No. 2025JGZZ40), the Urban and Rural Construction Development Foundation of Jiangsu Province, and the Jiangsu Provincial Qinglan Project. The authors would like to thank Prof. Xuemin Sherman Shen of the University of Waterloo for his invitation and valuable suggestions on this book. The authors would also like to express their gratitude to the other researchers in the MMC Research Group for their contributions to the presented research works. In particular, Prof. Xin Wei respectfully expresses most sincere gratitude to his late mother, Ms. Aihua Li. The publication of this book is indebted to her silent support and encouragement.

Contents

Chapter 1
Introduction

1.1 Background

As the demand for comprehensive digitization and intelligence continues to grow, multi-modal services that integrate audio, visual, and haptic elements have emerged [1, 2]. The International Telecommunication Union (ITU) Network 2030 Focus Group has identified these services as killer applications for the B5G and 6G era due to their ability to enhance immersive experiences in scenarios such as industrial manufacturing, rehabilitation medicine, and intelligent education. Cisco's white paper on future network trends projects a 100-fold increase in global mobile data traffic by 2030, with 81% attributed to multi-modal services [3].

The rapid growth of multi-modal services has led to a massive increase in multi-modal data, including audio, visual, and haptic signals, posing significant challenges to current multimedia communication systems. In the IMT-2030 (6G) white paper published by the China Academy of Information and Communications Technology, it is highlighted that for fully immersive real-time interaction, transmission delay must be under 1 ms, data rate should be over 10 Gbps, and reliability should reach 99.99999% [4]. However, existing 5G-based systems have 20 ms end-to-end delay and a packet loss rate of around 10^{-5} with high deployment costs. Therefore, communication technology should be re-designed to fulfill the increase in multi-modal data volume and the requirements of humans' immersive experiences.

In the history of communication technology development, audio-visual communications have been the primary support and basis for the majority of current multimedia services. During audio-visual communications, cameras and microphones at the sender capture ultra-high definition content. Then, efficient audio and video coding are performed to compress this original content. Subsequently, during transmission, resources are allocated to audio-visual streams according to specific transmission conditions. Ultimately, high-resolution displays or virtual reality headsets at the receiver deliver premium audio-visual experiences for humans. Thus, techniques about audio-visual-related coding, transmission, reception, and rendering are crucial.

X. Wei et al., *Cross-Modal Communication Technology*, Wireless Networks,
https://doi.org/10.1007/978-3-032-00957-9_1

Take video communication as an example. In video coding, the latest international standard is the versatile video coding (VVC) standard [5]. Compared to the high efficiency video coding (HEVC), VVC reduces bitrates by nearly 50% at the same visual quality. It can also offer more flexibility for various video types and applications. In video transmission, dynamic adaptive streaming over HTTP (DASH), developed by 3GPP and MPEG, is one of the most widely used visual streaming standards [6]. Moreover, the rapid development of mobile video services challenges cellular networks. To address issues such as limited storage, varying computing capabilities, dynamic communication environments, random network establishment, and diverse services for large-scale video applications, techniques like device-to-device (D2D) communications [7], uncoded linear transform transmission [8] are often considered. In video reception and rendering, studies focus on enhancing the key indicators such as peak signal-to-noise ratio (PSNR), quality of service (QoS), and quality of experience (QoE) [9].

Although audio-visual communications can create a realistic presence, it falls short in supporting the physical interactions crucial for multi-modal services. To enable physical interaction between human and machine, especially haptic perception and feedback in remote scenarios, haptic communications have gained much attention in recent years [10]. Unlike audio-visual communications, haptic-enabled teleoperation systems require bidirectional exchange of haptic signals over wireless networks, forming a global control loop between human operators and actuators/remote devices. Communication delay is highly sensitive to the quality and stability of these systems. To meet the ultra-low latency and ultra-high reliability requirements, new infrastructures like the tactile internet have been developed. The IEEE Tactile Internet Standard Working Group (IEEE 1918.1) has done pioneering work in this field [11].

Current haptic communication-related studies focus on haptic coding, transmission, reception, and rendering. In haptic coding, approaches based on perceptual dead zones and predictive coding are used. Notably, the PVC-SLP method in [12], which uses acceleration-sensitive functions, has been selected as the tactile codec for IEEE 1918.1. For haptic transmission, research focuses on edge computing-based resource offloading, QoS-oriented transmission scheduling, and low-latency and high-reliability wireless access. For example, [13] proposes an efficient framework. Within it, humans can share connections and act as access points to meet the strict requirement of real-time haptic feedback and control transmission within $1ms$. In haptic reception and rendering, the goal is to develop accurate, real time, and miniaturized haptic display devices, such as gloves, computer mice, and enhanced haptic perception loops [14]. Additionally, the quality of received haptic signals is crucial, leading to the proposal of evaluation metrics like structural similarity index measure (SSIM) and spectral temporal similarity (ST-SIM) based on human perception [15].

1.2 Motivation and Necessity

As we know, audio-visual communications and haptic communications can handle their respective information types. However, in multi-modal services with coexisting audio, visual, and haptic signals, they fall short due to significant modality differences. Haptic streaming needs communication techniques with high-reliability and low-latency properties, while bandwidth-heavy audio-visual streaming requires enhanced mobile broadband-related techniques.

To address these issues and support multi-modal services, cross-modal communications, which integrate audio-visual and haptic communications, emerge as a promising paradigm [16, 17]. It systematically considers the transmission requirements and structural properties of signals with different modalities, breaks modality barriers, and leverages their potential correlations (or named as "semantics"). This paradigm involves the construction of communication architectures, the design of key techniques for coding, transmission, and signal recovery, as well as the development of related prototype systems. Its goal is to achieve low latency and high reliability for haptic streaming, as well as high throughput for audio-visual streaming. It also aims to enhance interactive and immersive experiences, offering a universal, generalizable, and scalable solution for multi-modal services.

Compared to audio-visual communications and haptic communications, the uniqueness of cross-modal communications is reflected in three aspects. The first is efficient compression. Guided by the rate-distortion theory, it explores potential correlations among different modality signals of the same object. Using these correlations as auxiliary information, a cross-modal coder can be designed for efficient compression of multi-modal source signals. The second is on-demand scheduling. It further explores temporal, spatial, and semantic associations among multi-modal streams to implement elastic scheduling and adaptive resource allocation. The aim is to ensure effective transmission of audio, visual, and haptic streams. The third is high-quality signal reception. By utilizing a cloud-edge-terminal framework, it employs intra-modality and inter-modality knowledge transfer for feature fusion and semantic enhancement. It also adopts deep learning-based models to accurately recover missing or distorted signals, guaranteeing reception quality and an immersive experience for humans.

1.3 Organization of This Book

From the above descriptions, the core idea of cross-modal communications aligns with the development and requirements of multi-modal services. However, designing its architectural framework and the related key techniques is challenging. On one hand, it is difficult to explore the correlations between audio, visual, and haptic signals, which are respectively crucial for coding and signal recovery both before and after transmission. On the other hand, the dynamic transmission environment and

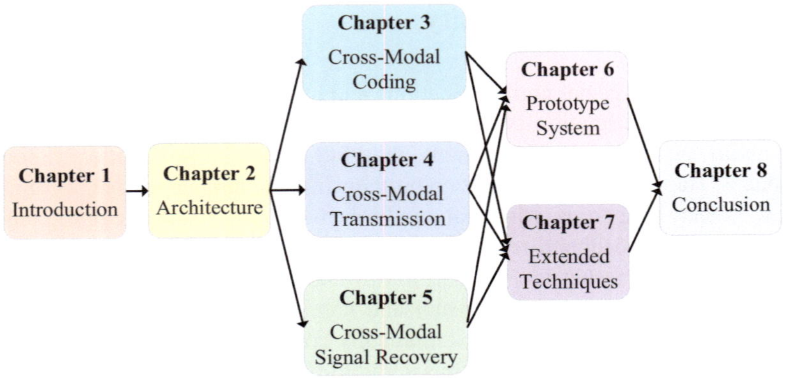

Fig. 1.1 Organization of this book

constrained resources restrict the design of transmission strategies. Despite significant progress in this domain, a book about cross-modal communication technology has yet to be presented. This research gap has prompted this book to provide a detailed description of this topic in response to the rapid development and pressing requirements of multi-modal services.

In this book, we investigate the cross-modal communication architecture, the related key techniques, and their applications in multi-modal services. Its organization is illustrated in Fig. 1.1. In this chapter, we present the background and motivation of this domain. In Chap. 2, we outline multi-modal services and a general cross-modal communication architecture, including typical service categories, communication requirements, architectural details, challenges, and design principles. In Chaps. 3, 4, and 5, we respectively propose three key techniques within the cross-modal communication architecture: cross-modal coding, cross-modal transmission, and cross-modal signal recovery. For each technique, we provide an in-depth presentation of the representative algorithms and the related experimental results. In Chap. 6, we describe an established database and the three developed cross-modal communication prototype systems. In Chap. 7, we introduce extended techniques, such as cross-modal semantic communications, cross-modal communications in extremely low-resource scenarios. Finally, we provide a summary of the entire book and discuss future research directions in Chap. 8.

References

1. Shen SX, Gao J, Li M, Zhou C, Hu S, He M, Zhuang W (2023) Toward immersive communications in 6G. Front Comput Sci 4:106848
2. Covaci A, Zou L, Tal I, Muntean G-M, Ghinea G (2019) Is multimedia multisensorial? a review of mulsemedia systems. ACM Comput Surv 51(5):91

3. Cisco Systems, Inc. (2024) The 2024 global networking trends report: uncover the critical strategies driving digital success
4. China Academy of Information and Communications Technology (2021) 6G overall vision and potential key technologies
5. Bross B, Wang Y, Ye Y, Liu S, Chen J, Sullivan GJ, Ohm JR (2021) Overview of the versatile video coding (VVC) standard and its applications. IEEE Trans Circ Syst Vid 31(10):3736–3764
6. Rodríguez DZ, Rosa RL, Alfaia EC, Abrahao JI, Bressan G (2016) Video quality metric for streaming service using DASH standard. IEEE Trans Broadcast 62(3):628–639
7. Zhou L, Wu D, Chen J, Dong Z (2018) Greening the smart cities: energy-efficient massive content delivery via D2D communications. IEEE Trans Ind Inform 14(4):1626–1634
8. Xiong R, Zhang J, Wu F, Xu J, Gao W (2017) Power distortion optimization for uncoded linear transformed transmission of images and videos. IEEE Trans Image Process 26(1):222–236
9. Wei X, Zhou L (2019) Multimedia QoE evaluation. Springer, Berlin
10. Antonakoglou K, Xu X, Steinbach E, Mahmoodi T, Dohler M (2018) Toward haptic communications over the 5G tactile internet. IEEE Commun Surv Tut 20(4):3034–3059
11. Holland O, Steinbach E, Prasad RV et al (2019) The IEEE 1918.1 "Tactile internet" standards working group and its standards. Proc IEEE 107(2):256–279
12. Hassen R, Gülecyüz B, Steinbach E (2020) PVC-SLP: perceptual vibrotactile-signal compression based-on sparse linear prediction. IEEE Trans Multimed 23:4455–4468
13. Mountaser G, Mahmoodi T, Simeone O (2018) Reliable and low-latency fronthaul for tactile internet applications. IEEE J Sel Areas Commun 36(11):2455–2463
14. Sundaram S, Kellnhofer P, Li Y, Zhu J, Torralba A, Matusik W (2019) Learning the signatures of the human grasp using a scalable tactile glove. Nature 569(7758):698
15. Hassen R, Steinbach E (2020) Subjective evaluation of the spectral temporal similarity (ST-SIM) measure for vibrotactile quality assessment. IEEE Trans Haptics 13(1):25–31
16. Zhou L, Wu D, Chen J, Wei X (2020) Cross-modal collaborative communications. IEEE Wirel Commun 27(2):112–117
17. Wei X, Zhou L (2021) AI-enabled cross-modal communications. IEEE Wirel Commun 28(4):182–189

Chapter 2
Cross-Modal Communication Architecture

Abstract In this chapter, the typical multi-modal services that have emerged in recent years are firstly described. These services can be categorized into three main types: haptic-dominant, audio-visual-dominant, and equally important multi-modal services. Following this, an analysis of the characteristics of different modalities is conducted. On the one hand, the heterogeneous communication requirements of audio, visual, and haptic signals are examined. On the other hand, the diverse communication requirements of the different categories of multi-modal services are also presented. Subsequently, to meet these requirements, a cross-modal communication architecture is constructed and the details of its components are described. Finally, the key techniques within this architecture are introduced. Additionally, the issues that need to be addressed as well as the core ideas behind the design of these key techniques are also discussed.

2.1 Typical Multi-Modal Service

Figure 2.1 lists typical multi-modal services. Based on the proportion and degree of integration of audio, visual, and haptic signals, these multi-modal services can be divided into the three main categories [1, 2]: Haptic-dominant multi-modal service, audio-visual-dominant multi-modal service, equally important multi-modal service.

2.1.1 Haptic-Dominant Multi-Modal Service

In haptic-dominant multi-modal services, haptic signals play a dominant role. Typical application scenarios include human-machine haptic interaction, rehabilitation training, and remote industrial robotic grasping, etc.

Human-machine haptic interaction: It is a highly promising research direction in the field of human-computer interaction [3]. Through the synergistic action of haptic sensors, actuators, and computational models, it enables the bidirectional transmission of haptic information, simulating human haptic perception and pro-

X. Wei et al., *Cross-Modal Communication Technology*, Wireless Networks,
https://doi.org/10.1007/978-3-032-00957-9_2

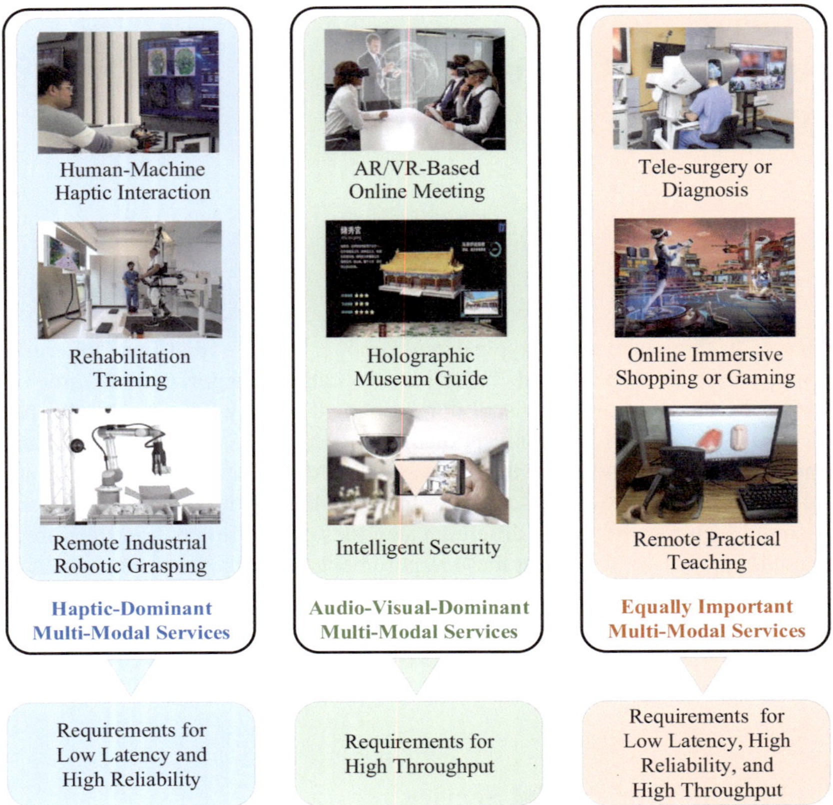

Fig. 2.1 Three categories of the typical multi-modal services

viding humans with a more natural, intuitive, and immersive interactive experience. Compared with traditional audio-visual interaction methods, haptic interaction has unique advantages and characteristics. (1) It can perceive a variety of physical properties such as pressure, vibration, temperature, and texture. It also can simulate the hardness, roughness, and vibration sensations when humans touch virtual objects, thereby enhancing their sense of immersion. (2) It allows humans to interact with devices more naturally, reducing learning cost and operational complexity. (3) It can generate corresponding haptic signals in real time based on humans' actions and environmental conditions, helping them to quickly adjust their operations, reduce errors, and achieve precise control.

Rehabilitation training: By simulating realistic haptic feedback, it provides patients with a more natural, intuitive, and personalized rehabilitation experience, significantly enhancing the effectiveness of rehabilitation and patient satisfaction. Specifically, (1) in upper limb functional rehabilitation applications, the use of haptic feedback gloves and upper limb rehabilitation robots can offer patients a variety of rehabilitation training modes, including active movement, passive movement, assisted movement, and resistive movement. These devices, by simulating realistic haptic feedback, help patients restore the motor functions of their hands and upper limbs, thereby improving their ability to perform self-care. (2) In fine hand movement rehabilitation applications, through "track-guided" tasks, patients can perform hand-eye coordination training in a virtual environment. Meanwhile, the system quantitatively assesses the patient's fine motor flexibility and stability based on the time taken to complete the task, contact force, and characteristics of the movement trajectory. (3) In the field of pediatric rehabilitation, by designing a diverse range of touch experiences, such as sensory integration training devices with haptic stimulation, they help children improve their sensory integration abilities and enhance their motor coordination and cognitive skills. (4) In remote and home-based rehabilitation applications, through remote monitoring and guidance, patients can use haptic feedback devices for rehabilitation training at home. At the same time, doctors can monitor the patient's training data in real time via the network and provide personalized rehabilitation recommendations.

Remote industrial robotic grasping: By integrating advanced sensor technologies, communication networks, and robotic control algorithms, remote robotic grasping systems are capable of performing high-precision and highly reliable operations in complex and dynamic industrial environments [4]. They are gradually becoming a key means of enhancing production efficiency, reducing costs, and ensuring operational safety. The characteristics of these systems include three main aspects. (1) The integration of force sensors and haptic sensors enables real-time feedback on the position, shape, weight, and surface properties of objects, allowing for the precise grasping of components of various shapes and sizes. (2) The systems possess flexibility and adaptability, enabling them to automatically adjust the grasping force and posture based on the shape and surface characteristics of the workpieces. (3) In hazardous or hard-to-reach industrial environments, such as underground pipeline maintenance and landslide rescue operations, high-definition cameras and sensors installed on robots allow operators to remotely control the robots from a safe environment, thereby ensuring operator safety.

2.1.2 Audio-Visual-Dominant Multi-Modal Service

In audio-visual-dominant multi-modal services, the importance of audio and visual signals is greater than that of haptic signals. Typical application scenarios include AR/VR-based online meeting, holographic museum guide, intelligent security, etc.

AR/VR-based online meeting: By providing a more natural, intuitive, and immersive interactive experience, it significantly enhances the efficiency and quality of remote collaboration [5]. Its characteristics include four main aspects. (1) By creating a virtual three-dimensional space, it makes humans feel as if they are in a real meeting environment, which not only enhances their engagement but also reduces the distractions commonly found in traditional video conferences. (2) Participants can operate the virtual interface through gestures, communicate through voice, and even express their intentions through body movements. This natural interaction mode improves the efficiency of the meeting. (3) It enables collaborative document editing in the virtual space, real-time viewing and discussion of 3D models, and even virtual presentations. This not only improves team collaborative efficiency but also promotes creative idea generation. (4) It supports a variety of devices and platforms, allowing participants to participate in meetings using VR headsets, AR glasses, smartphones, or tablets. Additionally, the system is scalable and supports large-scale remote collaboration.

Holographic museum guide: It is a novel guiding method that integrates holographic projection techniques with AR and VR [6]. By creating realistic three-dimensional virtual images and interactive experiences, it offers an immersive visiting experience to the visitors. This technology not only enhances visitor engagement and educational outcomes but also paves new avenues for cultural dissemination and museum education. Its characteristics include three main aspects: (1) Visitors can appreciate the exhibits from different angles without wearing any special devices, as if they are in a real historical scene. This immersive experience not only enhances the audience's participation but also improves their understanding of the exhibits and historical context. (2) Visitors can interact with the virtual exhibits, such as "touching" cultural relics, watching the restoration process of cultural relics, or participating in the recreation of historical scenes, thereby enhancing educational effectiveness. (3) Virtual guides can provide personalized interpretation services according to the visitor's needs.

Intelligent security: It establishes a comprehensive and intelligent security safeguard system through the deep integration of technologies such as AI, big data, and 5G. The main characteristics include three main aspects. (1) It implements real-time video analysis using AI algorithms, which can automatically detect abnormal behaviors (e.g., intrusion, fire, and congestion) and trigger alarms. This enhances response speed while reducing human intervention needs, which is particularly crucial in public places like airports and stations. (2) Based on big data analysis, it can predict potential risks and take preventive measures in advance. This predictive capability significantly improves the prevention of security incidents. (3) By leveraging high-speed communication technologies such as 5G, it enables remote real-time control and monitoring. Humans can view surveillance footage anytime via smartphones or computers to ensure safety.

2.1.3 Equally Important Multi-Modal Service

In equally important multi-modal services, both audio-visual and haptic signals play important roles. Typical application scenarios include tele-surgery or diagnosis, online immersive shopping or gaming, remote practical teaching, etc.

Tele-surgery or diagnosis: It integrates advanced communication networks, robotic technology, and multi-modal feedback systems to provide remote medical services to patients [7]. This can not only break geographical barriers but also enhance the accessibility and quality of medical services. The characteristics include four main aspects. (1) With high-precision visual feedback, doctors can precisely operate surgical instruments or make diagnostic conclusions even when the patient is hundreds or thousands of kilometers away. This is achieved through high-resolution 3D cameras and VR technology that provide a clear and real-time surgical view. (2) Through high-speed network connections and low-latency communication, doctors can interact with patients in real time during tele-diagnosis sessions, providing diagnostic recommendations and treatment guidance. During tele-surgery procedures, low latency ensures that the doctor's actions are immediately reflected in the surgical instruments, reducing the potential inaccuracies and risks associated with network delays. (3) By employing haptic feedback technology, the system simulates the haptic sensations between surgical instruments and tissues, allowing doctors to operate more naturally. This not only enhances surgical precision but also reduces surgical time and potential complications. (4) The use of robotic technology enables precise replication of the doctor's actions at the console. This enhances surgical safety by reducing physiological tremors and increasing the precision of the surgical instruments.

Online immersive shopping or gaming: It provides humans with an unprecedented experience by integrating audio, visual, and haptic signals [8]. In online immersive shopping, humans can virtually stroll through a digital store, employing sight, sound, and touch to gain a comprehensive understanding of product details, thereby making more informed purchasing decisions. In the immersive gaming, it creates an immersive game world for players through highly realistic visual effects, rich audio experiences, and haptic feedback. The incorporation of haptic signals, such as vibrations from motors that simulate in-game actions, further enhances player engagement. For instance, players can feel the vibrations associated with their character's running, jumping, or even combat actions, making the gaming experience more realistic and enjoyable through this multi-modal interaction. The characteristics of these services include three main aspects. (1) It emphasizes on multi-modal interaction, which involves the coordinated engagement of multiple senses, such as vision, hearing, and touch, to provide humans with a holistic experience. (2) It offers a high level of immersion, allowing humans to fully immerse themselves in the virtual environment. (3) Through analysis of shopping habits or behavioral preferences, it supports personalization, providing humans with tailored product recommendations or customized game content.

Remote practical teaching: By integrating audio, visual, and haptic signals, it offers students an immersive learning experience, which plays a significant role during the COVID-19 pandemic [9]. Specifically, it employs multi-modal interaction techniques to combine multiple sensory channels, including vision, auditory, and touch, enabling students to interact with virtual environments in various ways. For example, in a virtual laboratory, students can observe the experimental operation process through high-definition video. Meanwhile, they can use haptic feedback devices to simulate touch sensations during the experiment, such as the vibrations or pressures when operating tools. This multi-modal interaction not only enhances students' immersion but also increases the interest and effectiveness of learning. The characteristics of remote practical teaching include three main aspects. (1) It relies on efficient data transmission and processing techniques to ensure real-time and smooth performance. (2) It can adapt the difficulty and content based on students' performance in virtual experiments, ensuring that each student can learn at a pace suitable for themselves. (3) Based on the haptic feedback technique, it provides a quantitative evaluation of the standardization of students' practical operations, thus ensuring the effectiveness of practical teaching even when students are online.

2.2 Characteristics of Different Modalities

2.2.1 Communication Requirements for Signals

From the above descriptions about multi-modal services, it can be seen that they face the reality of coexistence signals with different modalities such as audio, visual, and haptic. When designing communication schemes, it is also necessary to consider transmission requirements of different modalities, shown in Table 2.1.

Maximum tolerable delay

Audio modality: The maximum tolerable delay for audio signals is $\leq$ 150 ms. This is because human perception of sound is highly sensitive. When the delay of an audio signal exceeds 150 ms, humans will noticeably feel that the sound is out of synchronization with actions or visual images, thereby affecting the naturalness and smoothness of communication.

Visual modality: The maximum tolerable delay for visual signals is $\leq$ 350 ms. Visual signals contain a large amount of visual frames, and their delay has a slightly lower impact on human experience compared to audio signals, but it still cannot exceed 350 ms. If this delay is too high, humans will feel a lag or delay in each visual frame, especially when watching real-time video conferences or immersive games. This kind of delay can seriously affect human participation and experience.

Haptic modality: The maximum tolerable delay for haptic signals is [1−60] ms. Haptic signals have the strictest delay requirements because haptic feedback needs

Table 2.1 Communication requirements for audio, visual, and haptic signals [1]

	Audio	Visual	Haptic
Delay (ms)	$\leq$150	$\leq$350	[1−60]
Jitter (ms)	$\leq$30	$\leq$30	[1−10]
Data loss rate (%)	$\sim$30	$\sim$20	$\sim 10^{-5}$
Data rate	[20−200] Kb/s	[2.5−40] Mb/s	$\sim$128 Kb/s
Closed loop	No	No	Yes

to be immediate. If the delay of a haptic signal is too high, humans will feel as if they are not touching the object or that the operation is not responsive, thereby reducing the sense of immersion.

Maximum tolerable jitter

Audio modality: Jitter refers to the variations in time intervals during signal transmission. Audio signals are highly sensitive to jitter because it can cause discontinuities or distortions in sound, thereby degrading the auditory experience.

Visual modality: Jitter in visual signals can affect display smoothness, leading to flickering or blurring of the images or videos. In video applications with real-time demands, such as AR/VR-based online meetings or holographic museum guides, excessive jitter can cause visual fatigue for humans and reduce the overall viewing experience.

Haptic modality: Haptic signals have even stricter requirements regarding jitter, as haptic feedback necessitates continuity and stability. If the jitter in haptic signals is too large, humans may perceive the surface of the object they are touching as uneven or unstable, which in turn can affect the accuracy and comfort of their operations.

Maximum tolerable data loss rate

Audio modality: Audio signals exhibit a certain degree of tolerance to packet loss because audio coding techniques can compensate for lost data through specific algorithms. However, when the packet loss rate exceeds 10%, the audio quality will significantly deteriorate, resulting in audio interruptions or distortions.

Visual modality: Visual signals contain a large amount of images or videos and thus have relatively higher tolerance to packet loss. Nevertheless, excessive packet loss can lead to mosaic effects or freezing in the video. In high-resolution visual transmission, controlling the packet loss rate is particularly important, as failure to do so can impair the visual experience.

Haptic modality: Haptic signals have extremely low tolerance to packet loss because haptic feedback requires a high degree of reliability and stability. In the transmission of haptic signals, even a minimal amount of data loss can cause users to perceive that they are not touching the object or that their operations have failed, thereby severely affecting the human experience.

Data rate

Audio modality: The bitstream requirement for audio signals is relatively low due to the high compression efficiency of audio data. However, high-quality audio signals may necessitate a higher bitstream.

Visual modality: Owing to the large amount of visual data, higher bandwidth is required for transmission. Different visual resolutions, frame rates, and encoding formats affect the size of the bitstream. For example, the bitstream for 1080 p video is typically around 510 Mb/s, while that for 4 K video may require higher bandwidth.

Haptic modality: The bitstream requirement for haptic signals is relatively stable because the structure of haptic feedback data is relatively simple, mainly comprising haptic and cutaneous information. Although this bitstream requirement is not high, the demands for real-time data transmission and reliability are extremely high.

Closed loop

Audio modality: Audio signals do not require a closed loop because audio transmission is typically unidirectional.

Visual modality: Visual signals also do not require a closed loop, as visual transmission is similarly unidirectional.

Haptic modality: Haptic signals require a closed loop because haptic feedback necessitates bidirectional transmission. For example, in a remote industrial robotic grasping scenario, the operator needs to receive haptic feedback from the robot through a haptic device while simultaneously sending operational commands to the robot. This bidirectional transmission requires a closed loop to ensure the synchronicity and consistency of the data.

2.2.2 Communication Requirements for Services

Table 2.2 presents a comparison of communication requirements for different categories of multi-modal services. Traditional multimedia services, such as audio and video calls, have relatively low bandwidth requirements, can tolerate delay greater than 100 ms, and have relatively relaxed reliability requirements. Rehabilitation training, as a haptic-dominant multi-modal service, requires the real-time transmission of high-definition visual and haptic feedback (for interaction with rehabilitation devices), thus demanding high bandwidth and low latency to ensure the synchronization of action commands and feedback. Holographic museum guides, which are primarily audio-visual-dominant services, have even higher bandwidth requirements for transmitting high-resolution holographic images. They also have stricter latency requirements to achieve real-time interaction between holographic projections and users, ensuring the continuity of the immersive experience. Tele-surgery or

Table 2.2 Comparison of communication requirements for different categories of multi-modal services

	Traditional multimedia service	Rehabilitation training	Holographic museum guide	Tele-surgery or diagnosis
Throughput	1–100 Mb/s	> 1 Gb/s	> 10 Gb/s	> 1 Gb/s
Delay (ms)	100–350	≤ 30	≤ 10	≤ 1
Reliability (%)	~95	~99.999	~99.999	~99.999

diagnosis, which has the highest requirements for latency and reliability, demands a bandwidth of >1 Gb/s, but the delay must be controlled within ≤ 1 ms, and the reliability must reach 99.999%. This ensures the absolute synchronization between the surgeon's operation of the instruments and the feedback from the patient's end. Any microsecond-level delay or packet loss could lead to surgical failure.

These differences in requirements reflect the varying degrees of integration of different modalities, such as audio, visual, and haptic, in multi-modal services. For instance, in tele-surgery or diagnosis, haptic signals account for up to 60% and require bidirectional real-time transmission of the haptic feedback from surgical instruments. In contrast, in holographic guides, visual signals account for over 80%, but their latency sensitivity is comparable to that of haptic signals. In summary, in haptic-dominant multi-modal services, the haptic modality has higher priority, while low-latency and high-reliability requirements should be given more attention. In audio-visual dominant multi-modal services, high throughput should be considered, while in equally important multi-modal services, latency, reliability, and throughput factors are all essential. This diversity of requirements means that the communication paradigm must simultaneously meet the demands of low latency, high reliability, and large capacity, and dynamically allocate resources for different services to achieve differentiated quality of service guarantees.

2.3 Communication Architecture

Based on the current audio-visual and haptic communications, traditional schemes for multi-modal services mainly transmit audio-visual and haptic streams through multiplexing techniques. One example is the use of adaptive application-layer multiplexing (ADMUX) [10]. It provides synchronization schemes between different modalities and achieves efficient dynamic bandwidth allocation through intelligent multiplexers. Another approach focuses on teleoperation systems with multi-modal feedback, which also belongs to an application-layer multiplexing approach [11]. In this approach, haptic signals are assigned the highest priority, and a preemptive recovery scheduling strategy is adopted for the transmission of audio-visual streams. The primary objective is to prevent latency and jitter during haptic signal reception while ensuring smooth audio-visual streaming. However, these works treat audio-visual streams and haptic streams separately, and give haptic modality higher priority. Due to the significant differences in characteristics and resource competition between the audio, visual, and haptic modalities, the availability and efficiency of these communication schemes are relatively low. Moreover, in many multi-modal services, the audio-visual signals dominate and are more important than the haptic signals. Therefore, it is inappropriate to always assign the haptic modality the highest priority in these cases.

Unlike previous communication schemes, an architecture of cross-modal communication, shown in Fig. 2.2, can better support multi-modal services [12]. This architecture mainly consists of three components: the master domain, the slave domain,

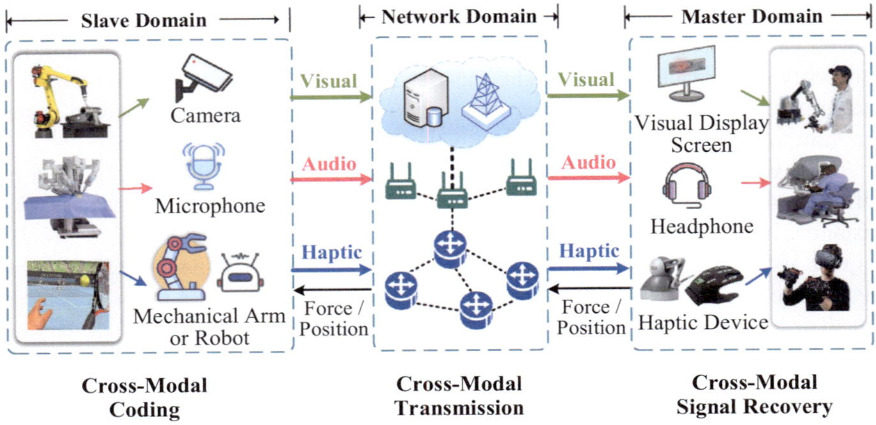

Fig. 2.2 An architecture of cross-modal communications [12]

and the network domain. It aims to break down barriers and fully utilize the potential correlations between audio, visual, and haptic signals. By meeting the requirements of low latency, high reliability, and high throughput, it seeks to enhance the quality of human-machine interaction and human immersive experience.

Master domain

It is responsible for receiving, processing, and rendering multi-modal signals. Operators send position-direction commands through the human-machine interface, while various devices integrated into the interface (e.g., haptic actuators, audio-visual display devices) convert the received signals into perceptible forms for the operators. Specifically, haptic actuators such as the Geomagic Touch and tactile gloves employ the advanced haptic feedback technology. The Geomagic Touch, based on the force feedback principle, accurately controls actuators like motors to simulate haptic properties such as object hardness and surface texture. This allows operators to experience realistic haptic resistance and feedback similar to physical operations during virtual manipulation. Tactile gloves use multiple miniature vibrators or electrical stimulators distributed across different parts of the glove to render object shapes, temperature changes, and vibration feedback. For example, in remote medicine practical teaching, they enable operators to sense tissue elasticity and friction between surgical instruments and tissues, greatly enhancing the realism and immersion of the operation. Audio-visual display devices such as PCs and mobile terminals can restore visual signals transmitted from the network domain into high definition, smooth videos using their powerful graphics processing capacities. They can also decode audio signals to restore realistic environmental sounds and voice prompts with their excellent audio processing capacities. For instance, in AR/VR-based online meetings, high-definition displays can clearly present real-time images of remote work sites, while surround sound speakers make operators feel as if they are on site, hearing sounds from all directions and enhancing the sense of presence. More importantly, the master domain must be equipped with a reception module to adapt to the multi-modal streams transmitted from the network domain. By employing the advanced signal decoding algorithms, it achieves fast and lossless decoding of multi-modal signals. In cases of packet loss or delayed arrival after transmission, cross-modal signal recovery techniques should be implemented to maximize the restoration of the multi-modal source signal, ensuring the reception quality and operator's immersive experience.

Slave domain

Serving as the source of multi-modal signals, it has acquisition and processing capabilities that directly determine the quality of communication. For haptic modalities, haptic perception sensors are usually installed on mechanical arms or robots. Specifically, haptic sensors are distributed along joints and end effectors to comprehensively sense forces and contact states in all directions during operation. For robots, haptic sensors are placed on body parts that frequently interact with the environment, such as limbs and the trunk, forming a haptic perception network. In general, sensors are closely integrated with the main body of the equipment, achieving miniaturization and low-power design through micro-electromechanical technology. This ensures stable and reliable signal acquisition without hindering normal equipment movement and operation, even in complex working environments. For visual modality, video acquisition devices such as high-resolution and high-frame-rate camera arrays are used to achieve comprehensive and detailed visual capture of the operational environment. For audio modality, auditory acquisition devices such as high-sensitivity, directional microphone arrays are employed. They can accurately pick up sounds in the environment, effectively suppress background noise, and employ sound source localization. Additionally, the slave domain must be equipped with a transmission module. Considering the characteristics of audio, visual, and haptic signals, it should extract key features and perform deep fusion. Based on network conditions and the importance of each modality in different multi-modal services, cross-modal coding techniques should be designed and implemented. This aims to minimize the data volume to be delivered while ensuring the retention of key features of each modality, achieving efficient compression.

Network domain

It serves as the link connecting the master domain and the slave domain, providing a wireless communication medium. The performance of its transmission strategy directly impacts the efficiency and stability of the entire cross-modal communication system. Specifically, cross-modal transmission techniques should be carefully designed. On the one hand, resource scheduling schemes need to be employed. These schemes dynamically adjust core indicators such as data rate according to network bandwidth, transmission environment, and other parameters to ensure transmission efficiency. On the other hand, differentiated resource allocation schemes can be implemented for audio, visual, and haptic streams with different priorities. Here, priority is given to ensuring the transmission quality of the critical modality stream, guaranteeing the stable and efficient operation of the cross-modal communication system.

Particularly, in this architecture, a cloud-edge-terminal collaborative framework can be adopted according to the demands of the specific task or scenario. Here, edge nodes are deployed near terminal devices in the master and slave domains to provide nearby services with the required computing, caching, and communication resources. Specifically, edge nodes located in the slave domain can be equipped with a transmission module. They are responsible for feature extraction, compression, coding, and packetization of signals collected by haptic sensors and audio-visual acquisition devices. Edge nodes located at the master domain can host a reception module to undertake functions such as cross-modal signal recovery. The edge nodes at both domains, along with the remote central cloud possessing robust computing and storage capabilities, collaborate to support the efficient implementation of cross-modal communication techniques.

2.4 Challenge and Design Principle

2.4.1 Challenge of Cross-Modal Coding

Due to the limitations of communication resources, multi-modal source signals should be compressed before transmission at the slave domain. Current coding schemes generally employ separate codecs for audio, visual, and haptic signals. For instance, VVC is used for video coding, and the method in [13] is applied for haptic coding. However, this approach significantly impacts the performance metrics of audio-visual and haptic streams, such as rate distortion. Thus, designing a cross-modal coding scheme is crucial for further compressing audio, visual, and haptic signals and eliminating redundancy among modalities. In the design of a cross-modal coding technique, three key issues need to be addressed.

(1) To overcome the insufficient coding efficiency and reliability limitations of traditional methods that ignore semantic redundancy, it is essential to fully exploit inter-modal semantic correlations for compressing multi-modal source signals. Thus, how to utilize correlations among different modalities becomes critical.
(2) Theoretically, it is necessary to investigate the potential benefits of cross-modal coding. For example, compared to traditional coding methods, how many fewer bits are required when using cross-modal coding to compress source signals? Moreover, under transmission constraints such as latency and reliability, what is the optimal coding rate achievable with cross-modal coding?
(3) How to design a general cross-modal codec based on signal processing and AI technologies. This codec should achieve efficient or even extreme compression while meeting transmission constraints and preserving key information integrity,

thereby supporting the terminal perception efficiency under the cross-modal communication architecture.

2.4.2 *Challenge of Cross-Modal Transmission*

Due to the significant differences in communication requirements among various modalities, it is necessary to achieve efficient multi-modal streaming. Specifically, haptic streams require low latency and high reliability during transmission, while audio-visual streams, with their relatively large capacity, need sufficient bandwidth to ensure data rates. Moreover, current wireless transmission resources and network capacity are limited, and the transmission environment is often dynamically changing. Therefore, for different categories of multi-modal services, the core and focus of corresponding transmission schemes should differ. Developing cross-modal transmission schemes to achieve efficient matching of transmission resources with transmission requirements is essential. In the design of cross-modal transmission techniques, three key issues need to be addressed:

(1) In different multi-modal services, there are differentiated contents and requirements. For example, in tele-surgery or diagnosis, the precision of haptic feedback is critical, while high-definition visual streams provide visual guidance for doctors, and both are indispensable. In online immersive gaming, real-time interactivity becomes the core requirement. These different categories of multi-modal services vary in the emphasis and importance of their transmission content. Identifying and meeting these differentiated key requirements is the primary challenge in designing specific cross-modal transmission schemes.
(2) The transmission requirements and priorities of audio-visual streams and haptic streams differ. Audio-visual streams usually have massive data volumes and extremely high bandwidth demands, while haptic streams require low latency and high reliability and are characterized by random and bursty arrivals. In some multi-modal services, haptic streams have the highest transmission priority. When haptic streams emerge, they compete with audio-visual streams for transmission resources, causing audio-visual streams to be delayed. This also leads to a significant drop in transmission efficiency and waste of transmission resources.
(3) Current wireless transmission resources and network capacity are limited and often dynamically changing. How to utilize caching, communication, computing, and other ways to dynamically allocate transmission resources under complex and ever-changing network conditions is important. In other words, how to achieve precise matching between multi-modal services and the requirements of each modality stream is a barrier that must be overcome in the design of cross-modal transmission techniques.

2.4.3 Challenge of Cross-Modal Signal Recovery

During cross-modal communications, signal loss and damage caused by transmission interference and noise pollution are inevitable. Particularly with the integration of haptic signals, their impairments can more easily degrade the human's immersive experience. Existing research solutions only consider signal recovery within a single modality, such as the reconstruction and compensation of audio and visual modality themselves [14]. Therefore, at the master domain of cross-modal communication architecture, it is crucial to establish a cross-modal signal recovery mechanism for the lost or damaged audio, visual, and haptic signals. This ensures the presentation quality of the received multi-modal signals and guarantees an immersive experience. In the design of the cross-modal signal recovery technique, three key issues need to be addressed.

(1) It is known that the received audio-visual and haptic signals have potential correlations, i.e., they belong to the same content or have common attributes. These correlations are the bonds connecting signals with different modalities. How to fully exploit and utilize these potential correlations to ensure the recovery quality of the desired modality signals is the first issue to be resolved.
(2) For different categories of multi-modal services, the emphasized modalities vary. For instance, for haptic-dominant multi-modal services like rehabilitation training, it is necessary to ensure the recovery quality of haptic signals and even to generate "virtual" touch sensations based on audio-visual signals. For equally important multi-modal services like online immersive shopping or gaming, visual and haptic signals are both prone to loss. Under these circumstances, algorithms with mutual recovery capabilities between haptic and visual modalities are required. Therefore, designing the corresponding cross-modal recovery mechanisms based on the reception quality requirements of the specific multi-modal service is also highly challenging.
(3) In addition to cross-modal signal recovery methods based on correlations among modalities, whether other approaches, such as retrieval and super-resolution reconstruction be utilized to achieve signal recovery and ensure an immersive experience.

2.4.4 Design Principle of Cross-Modal Communication Techniques

As evident from the above analysis, the potential correlation among audio, visual, and haptic modalities is of critical importance in designing cross-modal communication-related strategies. These correlations can be regarded as "semantics" independent of specific modalities [12]. In other words, data from diverse modalities, such as audio, visual, and haptic signals, despite varying forms, can convey identical semantics. Take a glass as an example: Its image, the sound of it being struck, and the touch sensation it provides are completely different, but all describe the semantic concept of a "glass". Semantics serves as a bridge across modalities, enabling information conversion and understanding between them. It allows communication systems to overcome modality differences and achieve unified concept or scene recognition.

In cross-modal communications, identifying and utilizing semantics that reflect correlations among modalities is the key to the successful design of cross-modal coding, transmission, and signal recovery techniques. Specifically, when designing a cross-modal coding scheme, semantic correlations can be used to compress one modality's content based on the others. In constructing cross-modal transmission strategies, analyzing the semantic importance of each modality stream can achieve efficient transmission resource matching with its requirements. Finally, in designing a cross-modal signal recovery mechanism, semantics can be tapped into to generate or restore the desired modality signals from undamaged ones.

In multi-modal services, semantics can be explored and obtained through three main ways. Firstly, meaningful semantic information can be annotated or incorporated into audio-visual and haptic streams, such as object category labels in the related visual and haptic signals. In this case, semantics can be derived in a supervised way by applying a specific learning algorithm. Secondly, when the annotation of meaningful semantic information is not possible, semantics need to be explored in an unsupervised way. For example, the delivered haptic packets and audio-visual packets from the same or different terminals contain important information. The received haptic packets and audio-visual packets may share the same or different synchronization time stamps, etc. These facts can help a specific learning algorithm identify the semantics. Thirdly, even if audio-visual and haptic signals at edge nodes or terminal devices lack exploitable semantics, useful information may be contained in other parts of the cross-modal communication architecture, i.e., at the central cloud. AI technology such as transfer learning and knowledge distillation can be used to acquire and transfer semantics from these sources.

2.5 Summary

In this chapter, we discuss three categories of typical multi-modal services and their characteristics. As these services involve audio, visual, and haptic modalities, their communication requirements are carefully analyzed, which should be given significant attention when constructing cross-modal architecture and the related key techniques within this framework. Different from existing frameworks for multi-modal services, the proposed architecture consists of three components, namely master domain, slave domain, and network domain, and can effectively support collaborative transmission and signal processing for multi-modal signals. Under this architecture, special attention should be paid to cross-modal coding, cross-modal transmission, and cross-modal signal recovery techniques. Finally, to address challenges in the design of these techniques, the concept of "semantics" should be deeply explored, analyzed, and leveraged, as this approach will bring significant benefits.

References

1. Zhou L, Wu D, Chen J, Wei X (2020) Cross-modal collaborative communications. IEEE Wirel Commun 27(2):112–117
2. Zhou L, Wu D, Wei X, Chen J (2021) Cross-modal stream scheduling for ehealth. IEEE J Sel Area Comm 39(2):426–437
3. Girbés-Juan V, Schettino V, Demiris Y, Tornero J (2023) Haptic and visual feedback assistance for dual-arm robot teleoperationin surface conditioning tasks. IEEE Trans Haptics 14(1):44–56
4. Darvish K, Penco L, Ramos J, Cisneros R, Pratt J, Yoshida E, Ivaldi S, Pucci D (2023) Teleoperation of humanoid robots: a survey. IEEE Trans Robot 39(3):1706–1727
5. Trinitatova D, Tsetserukou D (2023) Study of the effectiveness of a wearable haptic interface with cutaneous and vibrotactile feedback for VR-based teleoperation. IEEE Trans Haptics 16(4):463–469
6. Van der Hooft J, Amirpour H, Vega MT, Sanchez Y, Schatz R, Schierl T, Timmerer C (2023) A tutorial on immersive video delivery: from omnidirectional video to holography. IEEE Commun Surv Tut 25(2):1336–1375
7. Tai Y, Qian K, Huang X, Zhang J, Jan MA, Yu Z (2021) Intelligent intraoperative haptic-AR navigation for COVID-19 lung biopsy using deep hybrid model. IEEE Trans Ind Informat 17(9):6519–6527
8. Gaudina M, Zappi V, Brogni A, Caldwell DG (2012) Haptic, audio, and visual: multimodal distribution for interactive games. IEEE Trans Instrum Meas 61(11):3103–3111
9. Han Z, Tu Y, Huang C (2023) A framework for constructing a technology-enhanced education metaverse: learner engagement with human machine collaboration. IEEE Trans Learn Technol 16(6):1179–1189
10. Eid M, Cha J, Saddik AE (2020) Admux: an adaptive multiplexer for haptic-audio-visual data communication. IEEE Trans Instrum Meas 60(1):21–31
11. Cizmeci B, Xu X, Chaudhari R, Alt N, Steinbach E (2017) A multiplexing scheme for multimodal teleoperation. ACM Trans Multim Comput 13(2):21
12. Wei X, Wu D, Zhou L, Guizani M (2025) Cross-modal communication technology: a survey. Fundam Res 5(5):2256–2267
13. Steinbach E, Hirche S, Kammerl J, Vittorias I, Chaudhari R (2011) Haptic data compression and communication. IEEE Signal Proc Mag 28(1):87–96
14. Sterling A, Rewkowski N, Klazky RL, Lin MC (2019) Audio-material reconstruction for virtualized reality using a probabilistic damping model. IEEE Trans Vis Comput Gr 25(5):1855–1864

Chapter 3
Cross-Modal Coding

Abstract In this chapter, a visual-haptic mutual redundancy elimination scheme is firstly provided by considering mutual assistance between these two modalities. Then, an audio-haptic redundancy elimination mechanism is explored from the perspective of time-frequency masking. Subsequently, by exploring the benefit of potential correlation, a visual-aided haptic coding scheme is proposed to support cross-modal communications. From a theoretical perspective, the minimum number of bits required to compress haptic signals under the rate conditions of visual streams is explored. From a technical perspective, AI-empowered cross-modal prediction and channel coding is designed to support this scheme. Finally, to further achieve the optimum rate of multi-modal streams, a general cross-modal coding strategy is presented, which concerns the mutual assistance between visual and haptic modalities.

3.1 Visual-Haptic Mutual Redundancy Elimination

In this section, a visual-haptic mutual redundancy elimination scheme is proposed, shown in Fig. 3.1 [1]. The inspiration of this scheme is drawn from the perceptual correlation between human vision and touch. The aim is to eliminate redundancy in both visual and haptic-related contents before delivery.

Haptic-aided visual redundancy elimination (H2VRE)

In multi-modal services, the ongoing high-resolution visual streams are voluminous and demand substantial radio spectrum bandwidth. When faced with restricted transmission capabilities, there is a significant drop in the quality of the visual streaming. To address the redundancy elimination issue of visual streams, we leverage the human perceptual characteristics. As per the perceptual discount principle [2], the sensitivity of one sensory organ can diminish due to the influence of stimuli from the other sensory organs. For instance, in the context of online immersive gaming, the emergence of intense haptic stimuli leads to a decrease in the player's visual concentration. This implies that the player's need for high visual resolution can be lessened as they involuntarily shift some of their focus toward the sense of touch. Drawing inspiration from

X. Wei et al., *Cross-Modal Communication Technology*, Wireless Networks,
https://doi.org/10.1007/978-3-032-00957-9_3

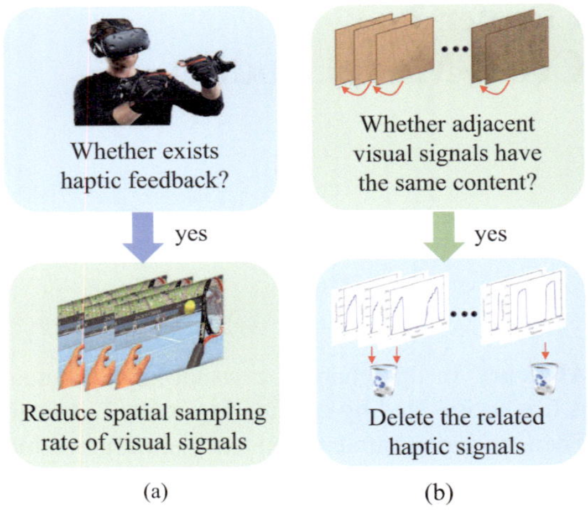

Fig. 3.1 Illustration of the visual-haptic mutual redundancy elimination: **a** Haptic-aided visual redundancy elimination; **b** Visual-aided haptic redundancy elimination

this phenomenon, the removal of redundancy in the visual modality can be realized with the assistance of haptic modality, as depicted in Fig. 3.1a. More specifically, before the transmission of visual streams, it is essential to check for the presence of haptic signals at each visual frame. If a haptic signal is detected, the spatial sampling frequency of the corresponding visual frame can be reduced to a suitable extent. If not, this spatial sampling frequency should stay the same.

Visual-aided haptic redundancy elimination (V2HRE)

In the context of haptic streams to be transmitted, these streams are marked by their latency sensitivity and bursty characteristics. To efficiently remove unnecessary haptic data, we resort to the perceptual interplay among various sensory channels once more. Take remote industrial robotic grasping as an example, the operator often cannot discern differences in touch sensations associated with similar visual representations such as identifying the surface material of an object. Inspired by this observation, the redundancy in haptic signals can be mitigated by utilizing visual modality, as illustrated in Fig. 3.1b. To be more specific, before the transmission of a haptic signal, the system initially assesses whether there is a minor difference between two successive haptic signals. When encountering similar haptic stimuli, it should further examine the associated visual frames. If the gap between these two corresponding visual frames is also minimal, it indicates that the haptic signals have not changed. In such cases, transmission is only required for the initial occurrence of the haptic signal, with the goal of minimizing the subsequent redundant transmissions. It should be added that the gap here means the labels or contents of the corresponding visual frames, according to the specific multi-modal services.

Besides considering redundancy elimination, it becomes crucial to meticulously craft the packet header prior to transmission. The packet header must encompass a field indicating the type of packet. This particular field is vital for identifying three distinct scenarios that arise from the H2VRE and V2HRE schemes mentioned above. In case 1, no sporadic haptic stimulus is detected and only the visual stream is active at a given moment. There is no reduction in spatial sampling frequency. The visual packet is marked with the standard visual (SV) type field for transmission. In case 2, when a bursty haptic stimulus is present at a particular time, activating both the H2VRE and V2HRE schemes, the visual packet is designated with the H2VRE-V2HRE type field. Conversely, the haptic packet does not need to be transmitted. In case 3, if a bursty haptic stimulus arises but does not meet the criteria to initiate the condition about V2HRE, only the H2VRE scheme is engaged for the visual stream. As a result, the visual packets are identified with the H2VRE type field, and the haptic packets are designated with the standard haptic (SH) type field for dispatch. Moreover, a time-stamp field must be incorporated. This time-stamp field links the visual packet with its corresponding haptic packet, both conveying the identical content at the origin. It aids in synchronization and guarantees the accurate sequence of rendering across different modalities upon reception.

The whole procedure of this visual-haptic mutual redundancy elimination scheme is described in Algorithm 3.1 [1].

Algorithm 3.1 Procedure of the visual-haptic mutual redundancy elimination

Require: Original visual frames $s^{(v)}$, original haptic frames $s^{(h)}$, ε, η
for $t = 1$ to T **do**
  if $s^{(h)}(t)$ appears at time t **then**
    Reduce sampling rate $R(s^{(v)}(t))$
    Detect the last time Δt that $s^{(h)}$ appears
    if $\left\| s^{(h)}(t) - s^{(h)}(t - \Delta t) \right\|_1 \leq \varepsilon$ **and** $\left\| s^{(v)}(t) - s^{(v)}(t - \Delta t) \right\|_1 \leq \eta$ **then**
      $s^{(h)}(t)$ does not need to be delivered
      Add a H2VRE-V2HRE type field into $s^{(v)}(t)$
    else
      Add a SH type field into $s^{(h)}(t)$
      Add a time-stamp field into $s^{(h)}(t)$
      Add a H2VRE type field into $s^{(v)}(t)$
    end if
  else
    Add a SV type field into $s^{(v)}(t)$
  end if
  Add a time-stamp field into $s^{(v)}(t)$
end for
Output: Visual packets $\tilde{s}^{(v)}$ and haptic packets $\tilde{s}^{(h)}$ after redundancy elimination

---

Experimental result and analysis

To evaluate the performance of the outlined H2VRE scheme, we simultaneously transmit continuous 4 K visual streams (with an average bitrate of 24 Mbps) and bursty haptic streams (with a sampling rate of 200 Hz). When haptic signals are present, the spatial sampling frequency of the visual frames is halved. We then contrast the throughput and latency of the H2VRE scheme against those of a scheme

Fig. 3.2 Performance of the H2VRE: **a** Throughput; **b** Latency

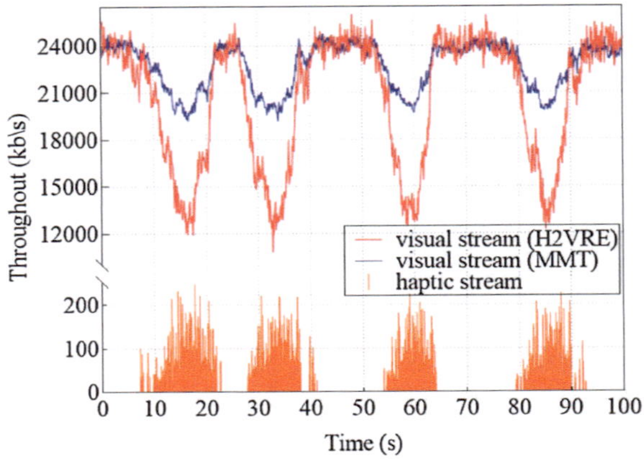

(a)

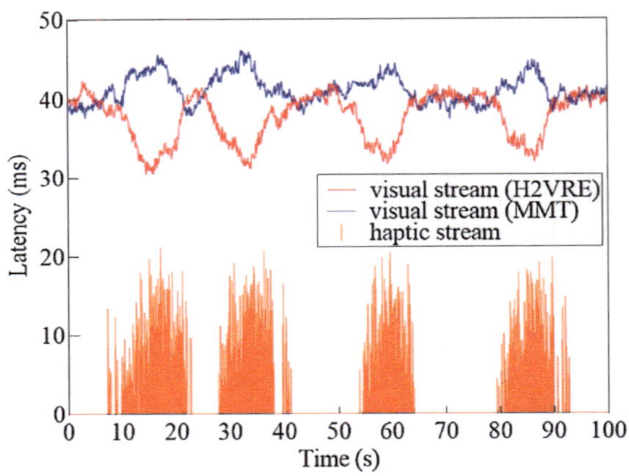

(b)

that does not incorporate H2VRE, such as the multiplexing multi-modal transmission scheme (MMT) [3]. As depicted in Fig. 3.2a, the implementation of H2VRE significantly reduces the throughput of the visual stream, facilitated by the haptic modality. This elimination reduction scheme can substantially free up or conserve resources, including transmission bandwidth and storage space, among others. Furthermore, the diminished size of the visual streams means they are less impacted by competition with haptic streams during transmission. Consequently, the transmission latency for the visual stream is notably reduced, as illustrated in Fig. 3.2b.

To evaluate the performance of the V2HRE scheme, we also simultaneously transmit visual streams (with a frame rate of 30 fps) and bursty haptic streams. Considering a decennial frame interval, if the variation in visual content between the current and the preceding interval is minimal, the haptic packet for the current interval is omitted.

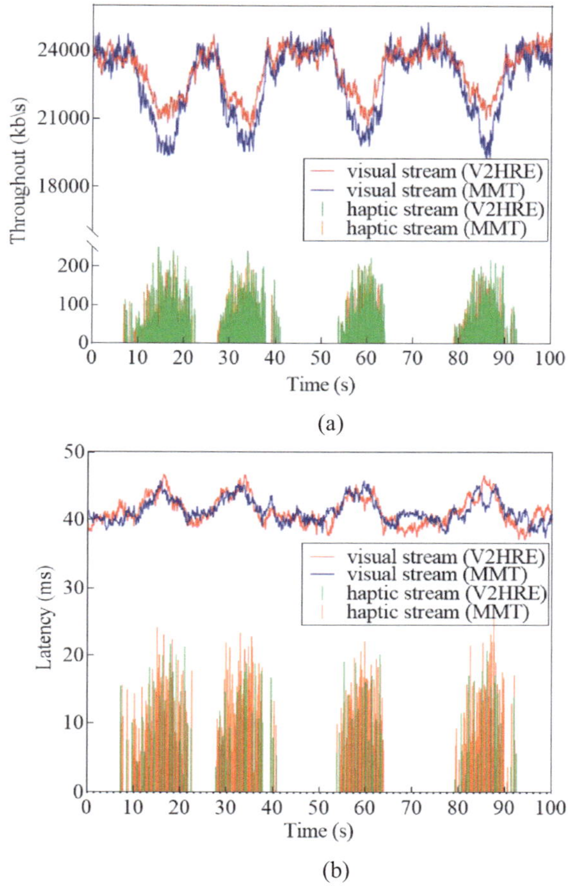

Fig. 3.3 Performance of the V2HRE: **a** Throughput; **b** Latency

We then contrast the throughput and latency of the V2HRE scheme against those of a scheme devoid of V2HRE (e.g., MMT [3]). As depicted in Fig. 3.3a and b, by eliminating redundancies in haptic streams, the competition among modalities is reduced. Consequently, the throughput and latency of visual streams are markedly improved. Additionally, the transmission efficiency of the haptic stream itself is also enhanced.

3.2 Audio-Haptic Joint Redundancy Elimination

As two representative non-visual senses, auditory and touch perceptions share numerous commonalities. On one hand, the size of audio and haptic signals is relatively small and they have analogous waveforms in both time and frequency domains. On the other hand, humans experience similar masking effects within their auditory and touch sensation systems [4]. The former observation prompts us to consider the integration of audio and haptic signals. The latter suggests that we might lower the intensity of these two non-visual signals. In essence, it could be beneficial to assess and eliminate the redundancy between these non-visual signals collectively, potentially enhancing the efficiency of visual streaming at the source. Therefore, in this section, a time-frequency masking-based audio-haptic redundancy elimination mechanism has been designed [5].

The masking effect is a principle derived from human perception theories, aimed at eliminating redundant sensory information without affecting the integrity of the original signals. To be more specific, it indicates that dominant frequency components in audio or haptic signals can overshadow those with less intensity. Consequently, the less significant, low-frequency components that are not crucial for transmission can be effectively filtered out. While frequency masking has been thoroughly explored in practical scenarios, the application of time masking, especially in a quantitative sense, is not as common. It's important to note that time masking, particularly backward time masking, should not be disregarded due to its extended duration and substantial masking effect, which plays a key role in compression. Thus, it is imperative to take into account both frequency and time masking in a comprehensive manner.

Motivated by this concept, the mechanism unfolds within the time-frequency domain (shown in Fig. 3.4), allowing for the simultaneous capture of both global and local time and frequency details. Here, the masking threshold is depicted by a two-dimensional surface, with the red solid circle indicating the masking signal or unit. Purple solid circles representing signals with amplitudes below the threshold, denoted as the masked unit, are candidates for masking. Orange circles, with amplitudes exceeding the threshold, are chosen for transmission. Figure 3.5 provides the specific realization procedure of the proposed audio-haptic joint redundancy elimination (AHJRE). Initially, audio and haptic signals are converted into time-frequency

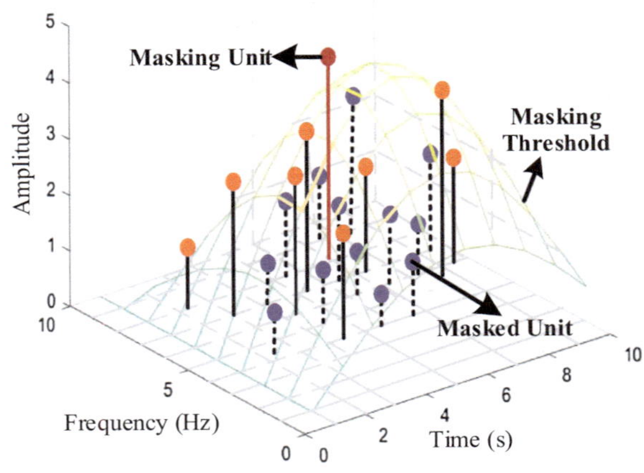

Fig. 3.4 Illustration of masking effect in time-frequency domains

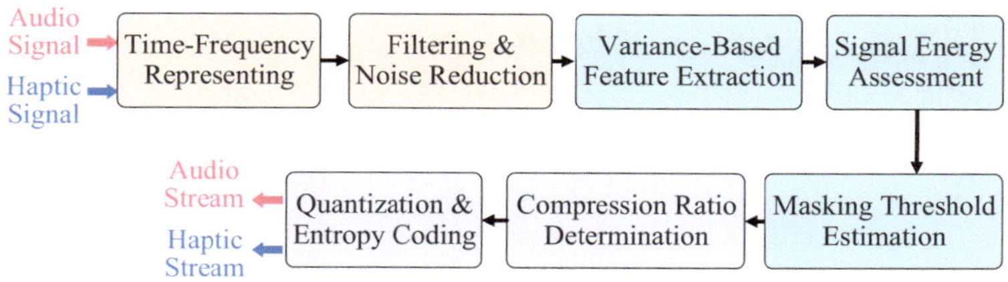

Fig. 3.5 Procedure of the audio-haptic joint redundancy elimination

representations using wavelet transforms, offering richer features than those obtained from time or frequency domains alone. The resulting time-frequency plane undergoes filtering, noise reduction, and is then used to extract features based on variance, with signal energy being assessed for each time-frequency unit. Subsequently, an unsupervised, nonparametric adaptive method is employed to estimate the masking threshold from the variance and energy [6], taking into account the current signal's preceding and subsequent frames. It's important to note that this threshold adapts dynamically to the specific characteristics of the audio and haptic signals. After the masking threshold is derived, the signal masking ratio or compression ratio is determined. This ratio is finally applied across all time-frequency units for the purposes of quantization and entropy encoding. If the audio and haptic features within the time-frequency domain fall below their respective masking thresholds, they are not transmitted. Consequently, by utilizing this procedure, the redundancy of audio and haptic signals can be significantly reduced or compressed without any noticeable difference in perception.

Experimental result and analysis

Figure 3.6 depicts the performance of this redundancy elimination scheme. Initially, Fig. 3.6a presents the average throughput of the multi-modal streams over a certain period of time. On one hand, the transmission efficiency for audio and haptic streams is enhanced due to the elimination of their redundancies. On the other hand, the compression of these two modality streams mitigates the resource competition with the visual stream. This results in an increased throughput for the visual stream as compared to the MMT scheme [3]. Furthermore, Fig. 3.6b clearly demonstrates that the perceived quality of the compressed audio and haptic signals gradually degrades with the increasing compression ratios. Nonetheless, the redundancy elimination mechanism proposed here outperforms the MMT. This improvement is attributed to the utilization of time-frequency masking effects for both audio and haptic modalities, which allows for simultaneous optimization in cross-modal communications.

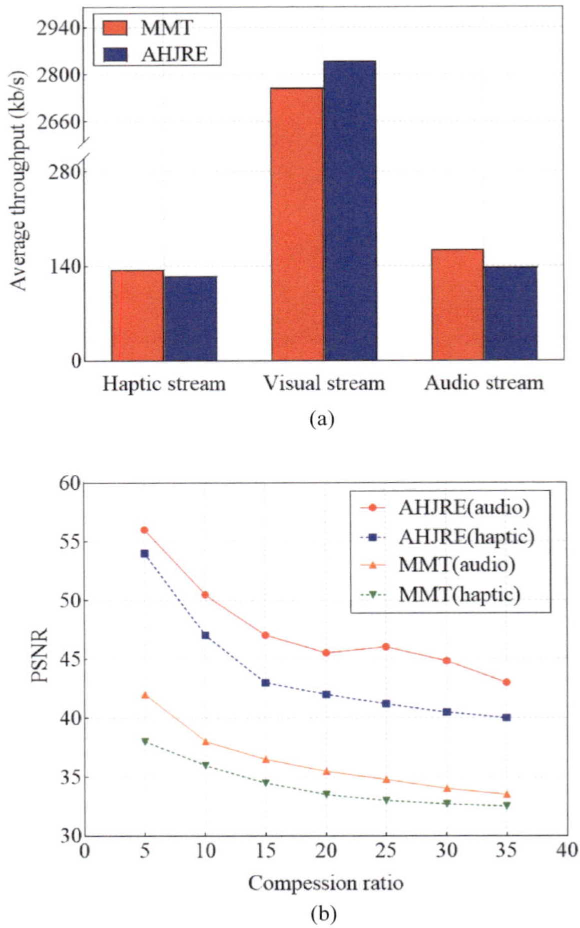

Fig. 3.6 **a** Average throughput; **b** PSNR of the proposed redundancy elimination mechanism and that in [3]

3.3 Visual-Aided Haptic Coding

Traditional multi-modal coding schemes usually design codecs for haptic, visual, and audio modalities separately [7, 8]. Obviously, this cannot handle multi-modal source signals effectively. Specifically, in haptic-dominant multi-modal services, when the haptic signals emerge, the rate-distortion performance of the visual stream is severely affected. The main reason is that, to meet the low-latency and high-reliability requirements of haptic streams, haptic modality is usually given higher communication priority than visual modality [3]. Under such circumstances, due to competition for limited communication resources, random and unpredictable frequent requests for touch will inevitably affect the stable visual streaming. Therefore, a natural idea is to transmit a very small amount of haptic information by compressing the haptic source signals as much as possible, thereby reducing their impact on visual streaming. Currently, it is possible to predict haptic signals from visual signals by designing suitable cross-modal signal recovery schemes (described in Sect. 3.5). Considering this, a hypothesis is proposed: A coding scheme can be designed to reduce the bitrate of the haptic stream through the potential correlation or semantics between haptic and visual modalities [9]. The details of this coding scheme will be described in the following.

3.3.1 Cross-Modal Coding Model and Semantic Entropy

Semantic ambiguity is a challenge firstly to be handled when leveraging potential correlations. Existing coding schemes are based on classical information theory. They only describe what source symbols the sender chooses, but ignore the semantics the sender intends to express [10]. Therefore, it is necessary to investigate a coding model and information metric theory to describe the symbols and semantics of multi-modal sources.

Firstly, a cross-modal coding (CMC) model is constructed, shown in Fig. 3.7. The encoders observe a pair of haptic source signal S_h and visual source signal S_v. They takes values in $\mathcal{S}_h \times \mathcal{S}_v$, which are drawn i.i.d. from probability mass function $p_{S_h,S_v}(s_h, s_v)$. The semantic properties of S_h and S_v are denoted as P_h and P_v, respectively. P_h and P_v belong to a common semantic property set $\mathcal{P}$. Based on this, $P_h - P_v$ is defined as the semantic relationship between P_h and P_v. If $P_h - P_v = 0$, S_h is semantic correlated with S_v. For example, although a pair of haptic-visual source signals depicts the characteristics of stone from different perspectives, the same semantic "stone" can be extracted ($P_h = P_v =$ "stone"). To be more specific, two sequences $S_h^n = \{S_{h,i}\}_1^n$ and $S_v^n = \{S_{v,i}\}_1^n$ are respectively represented as each haptic and visual frames, where n is the number of samples/blocks in per frame.

To implement visual-aided haptic coding, a haptic frame S_h^n, its corresponding visual frame S_v^n, and their semantic relationship $P_h - P_v$ should be encoded. The reason is that only if the decoder knows $P_h - P_v$, it can find the semantically correlated visual blocks and reconstruct haptic samples assisted with them. Here, we assume that $\{S_h, P_h - P_v\}$ is encoded with R_h bits, while S_v is encoded with R_v bits.

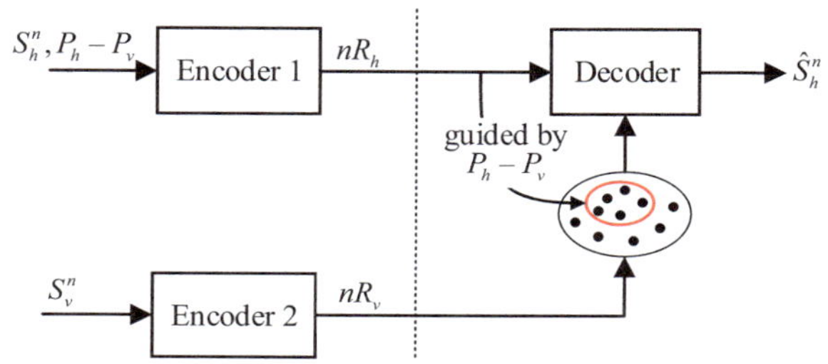

Fig. 3.7 Illustration of the CMC model. The black dots in the red line represent the received visual blocks which are semantically correlated with the haptic frame S_h^n [9]

Moreover, the decoded haptic samples are denoted as $\hat{S}_h$ also taking values in $\hat{\mathcal{S}}_h$, with the distortion level denoted as D. Based on the above notations, the mathematical expression of the CMC model is defined in the following.

CMC model: An $(n, 2^{nR_h}, 2^{nR_v})$ cross-modal code for the haptic-visual source pair (S_h, S_v) comprises two encoding functions,

$$\begin{aligned} f_h &: \mathcal{S}_h^n \times \mathcal{P} \rightarrow \left\{1, 2, \ldots, 2^{nR_h}\right\}, \\ f_v &: \mathcal{S}_v^n \rightarrow \left\{1, 2, \ldots, 2^{nR_v}\right\}. \end{aligned} \tag{3.1}$$

and a decoding function,

$$g_h : \left\{1, 2, \ldots, 2^{nR_h}\right\} \times \left\{1, 2, \ldots, 2^{nR_v}\right\} \rightarrow \hat{\mathcal{S}}_h^n. \tag{3.2}$$

Moreover, to conveniently verify the CMC model, the baseline model [11] should also be given.

Baseline model: An $(n, 2^{nR_h'})$ code for S_h contains a encoding function $f_h : \mathcal{S}_h^n \rightarrow \{1, 2, \ldots, 2^{nR_h'}\}$, and a decoding function $g_h : \{1, 2, \ldots, 2^{nR_h'}\} \rightarrow \hat{\mathcal{S}}_h^n$. Here R_h' is the number of bits required to describe S_h. There are two different tolerance levels for haptic signals distortion at the receiver: If $D = 0$, $R_h' \geq H_r(S_h)$; If $D > 0$, $R_h' \geq I_r(S_h; \hat{S}_h)$.

Next, under the constructed CMC model, a quantity which can measure "random" and "fuzzy" nature of semantic uncertainty within it should be defined. This quantity will guide us to compute how many bits are required to describe a pair of multi-modal sources' symbols and potential correlations (or semantics). Referring to the definition of Shannon entropy [10], a notion about *semantic entropy* is presented, regarded as the extension of Shannon entropy at the semantic level. Specifically, for any given discrete source S, its semantic entropy $H_s(S)$ enjoying the properties $\mathcal{P}$ can be defined:

$$H_s(S) = \underbrace{-\sum_{s \in \mathcal{S}} p_S(s) \log\, p_S(s)}_{random} \underbrace{- \sum_{P \in \mathcal{P}} \mathcal{F}_P(S) \log \mathcal{F}_P(S)}_{fuzzy}. \tag{3.3}$$

The first item in Eq. (3.3) is "random" uncertainty. It is related to the freedom of choice when a source selects its value in the alphabet. Take the haptic modality as an example, the "random" nature of S_h can be measured by Shannon entropy,

$$H_r(S_h) = -\sum_{s_h \in \mathcal{S}_h} p_{S_h}(s_h) \log\, p_{S_h}(s_h). \tag{3.4}$$

The second item in Eq. (3.3) is "fuzzy" uncertainty. It concerns the degree of "fuzziness" whether a source enjoys a property. For the haptic modality, it is

$$H_f(S_h) = -\sum_{P\in\mathcal{P}} \mathcal{F}_P(S_h)\log\mathcal{F}_P(S_h). \tag{3.5}$$

Here, a fuzzy set $\mathcal{F}_P$ on n-length sequence S^n is a function $\mathcal{F}_P : S^n \to [0, 1],\ P \in \mathcal{P}$. $\mathcal{F}_P(S)$ is a numerical evaluation of the "degree" to which source S enjoys the property P. Therefore, $\{\mathcal{F}_{h,P}\}_{P\in\mathcal{P}}$ is used to denote M fuzzy sets defined in $\mathcal{S}_h$. When $\mathcal{F}_{h,P}(S_h) = 1/2$, it means whether S_h enjoys P cannot be determined. In this case, the value of $H_f(S_h)$ is the largest. On the contrary, when $\mathcal{F}_{h,P}(S_h)$ is near 0 or 1, $H_f(S_h)$ becomes small.

In order to further explain Eq. (3.3), a simple example is given. We assume that $S = 1$ with probability p and $S = 0$ with probability $1 - p$. S enjoys property P_0 with fuzzy functions $\mathcal{F}_{P_0}(S) = q$ and $\mathcal{F}_{P_1}(S) = 1 - q$. In this case, the semantic entropy is

$$\begin{aligned} H_s(S) &= p\log 1/p + (1-p)\log 1/(1-p) \\ &\quad + q\log 1/q + (1-q)\log 1/(1-q) \\ &= H_s(p, q). \end{aligned} \tag{3.6}$$

Moreover, when $q = 0$ or 1, $H_s(p, q)$ achieves lower bound. This is because S enjoys the property P_0 at $q = 0$, while S enjoys the property P_1 at $q = 1$. Under this condition, S is not fuzzy. The semantic entropy is equivalent to the Shannon entropy, $H_s(p, q) = H_r(p)$. When $q = 1/2$, $H_s(p, q)$ gets the upper bound. It means the "degree" to which S enjoys the property P is $1/2$. Under this condition, the uncertainty of the "fuzzy" nature is maximum.

Based on the definition of original semantic entropy in Eq. (3.3), for the pair of S_h and S_v, the *joint semantic entropy* with the property $\mathcal{P}$ and a joint distribution $p_{S_h,S_v}(s_h, s_v)$ is defined as

$$\begin{aligned} H_s(S_h, S_v) = &-\sum_{\substack{s_h\in\mathcal{S}_h,\\ s_v\in\mathcal{S}_v}} p_{S_h,S_v}(s_h, s_v)\log\ p_{S_h,S_v}(s_h, s_v) \\ &-\sum_{P_h,P_v\in\mathcal{P}} \mathcal{F}_{P_h,P_v}(S_h, S_v)\log\mathcal{F}_{P_h,P_v}(S_h, S_v), \end{aligned} \tag{3.7}$$

where $\mathcal{F}_{P_h,P_v}(S_h, S_v)$ is the joint membership function representing the "degree" to which S_h, S_v with the properties P_h, P_v, respectively.

It can be seen that $H_s(S_h, S_v) \le H_s(S_h) + H_s(S_v)$ with equality only if S_h, S_v are semantic independent. In this case, $p_{S_h,S_v}(s_h, s_v) = p_{S_h}(s_h)p_{S_v}(s_v)$ and $\mathcal{F}_{P_h,P_v}(S_h, S_v) = \mathcal{F}_{P_h}(S_h)\mathcal{F}_{P_v}(S_v)$. If S_h, S_v are semantic dependent, the *conditional semantic entropy* of S_h given S_v, $H_s(S_h|S_v)$, can also be derived.

$$H_s(S_h|S_v) = -\sum_{\substack{s_h \in \mathcal{S}_h, \\ s_v \in \mathcal{S}_v}} p_{S_h,S_v}(s_h, s_v) \log\ p_{S_h|S_v}(s_h|s_v) - \sum_{P_h, P_v \in \mathcal{P}} \mathcal{F}_{P_h,P_v}(S_h, S_v) \log \mathcal{F}_{P_h}(S_h|S_v). \quad (3.8)$$

Therefore, the chain rule that existed in the Shannon entropy still applies between the joint and conditional semantic entropies.

$$H_s(S_h, S_v) = H_s(S_v) + H_s(S_h|S_v). \quad (3.9)$$

Similarly, the semantic entropy of a sequence $S^n = \{S_i\}_1^n$ satisfies

$$H_s(S^n) = \sum_{i=1}^{n} H_s(S_i|S_{i-1}, \ldots, S_1). \quad (3.10)$$

Finally, the *mutual semantic information*, which is a quantitative characterization of semantic correlation among modalities, can be defined

$$I_s(S_h, S_v) = \sum_{\substack{s_h \in \mathcal{S}_h, \\ s_v \in \mathcal{S}_v}} p_{S_h,S_v}(s_h, s_v) \log \frac{p_{S_h,S_v}(s_h, s_v)}{p_{S_h}(s_h) p_{S_v}(s_v)} + \sum_{P_h, P_v \in \mathcal{P}} \mathcal{F}_{P_h,P_v}(S_h, S_v) \log \frac{\mathcal{F}_{P_h,P_v}(S_h, S_v)}{\mathcal{F}_{P_h}(S_h) \mathcal{F}_{P_v}(S_v)}. \quad (3.11)$$

In other words, it describes the amount of semantics that one source contains about another source. Therefore, the relationship between semantic entropy and semantic mutual information is

$$\begin{aligned} I_s(S_h, S_v) &= H_s(S_h) - H_s(S_h|S_v) \\ &= H_s(S_v) - H_s(S_v|S_h) \\ &= H_s(S_h) + H_s(S_v) - H_s(S_h, S_v). \end{aligned} \quad (3.12)$$

3.3.2 Coding Theory Under Two Distortion Conditions

After constructing the CMC model and the related quantities, it should investigate how much benefit is derived from the semantic correlation among multi-modal source signals. Consequently, it is essential to establish the theoretical limits of this benefit (e.g., in terms of reducing the bit amount required to compress an information source) to guide the design of cross-modal codecs. In this section, the minimum number of bits required to describe the haptic source S_h aided by the visual source S_v is explored.

Specifically, the relationships between R_h, R_v, S_h and S_v are constructed. Based on this, the minimum number of bits R_h under the condition of R_v can be obtained.

According to the acceptable distortion threshold for haptic signals at the receiver (e.g., master domain in Fig. 2.2), the research about this benefit can be further divided into two main parts. On hand, the question "how many bits can be saved for haptic streams with the aid of visual modality and without causing any distortion ($D = 0$)" should be handled. On the other hand, the question "how many bits can be saved for haptic streams with the aid of visual modality and allowing several distortions ($D > 0$)" should also be tackled.

(1) To handle the first question ($D = 0$), a theorem about the characterization of R_h is proposed. Let a pair of sources (S_h, S_v) be $i.i.d. \sim p_{S_h,S_v}(s_h, s_v)$ and enjoys M properties $\mathcal{P} = \{P_j\}_1^M$. When $R_v \geq I_r(S_v; U)$, S_h can be decoded perfectly with cross-modal coding, if and only if

$$R_h \geq H_s(S_h|U), \tag{3.13}$$

for the joint probability mass function $p_{S_h,S_v}(s_h, s_v)p_{U|S_v}(u|s_v)$, where $|\mathcal{U}| \leq |\mathcal{Y}| + 2$. The proof of this theorem can be found in [9].

The auxiliary source U is the reconstruction of S_v with the transition probability $p_{U|S_v}(u|s_v)$ at the decoder. For the special case where $U = S_v$, $R \geq H_r(S_v)$, and thus S_v can be reconstructed perfectly. In this case, $R_h = H_r(S_h|S_v)$ bits suffice to describe S_h. At the other extreme, if $U = \emptyset$, S_h, $R_h = H_r(S_h)$ bits are necessary to describe S_h. This is also consistent with the result for the baseline model, i.e., $R_h{}' \geq H_r(S_h)$.

From the above theorem, if $\exists S_h{}' \in \mathcal{S}_h$, $\forall P \in \mathcal{P}$, such that $p_{S_h{}'|U}(s_h{}'|u) = \{0, 1\}$, $0 < \mathcal{F}_P(S_h{}'|U) < 1$ and $0 < p_{S_h{}'|S_h}(s_h{}'|s_h) < 1$, $\mathcal{F}_P(S_h{}'|S_h) = \{0, 1\}$, then R_h can be represented as

$$\begin{aligned} R_h \geq H_s(S_h|U) &= H_f(S_h|U) + H_r(S_h|U) \\ &= \underbrace{H_f(S_h|U)}_{\text{semantic relationship}} + \underbrace{H_r(S_h|S_h{}')}_{\text{symbol}}. \end{aligned} \tag{3.14}$$

Here, $H_f(S_h|U)$ is the minimum number of bits to describe the semantic relationship between S_h, S_v. $H_r(S_h|S_h{}')$ is the minimum number of bits to describe the semantic relationship between S_h and $S_h{}'$, where the $S_h{}'$ denotes the reference source of S_h based on U. It should be noted that $S_h{}'$ generated from U can be taken as a virtual noisy channel. The input of this channel is S_h, whose capacity is measured by the statistical correlation between $S_h{}'$ and S_h. Channel coding can be adopted to correct errors in $S_h{}'$ and reconstruct S_h from it.

(2) To tackle the second issue ($D > 0$), the rate-distortion function $R_h(D)$ should be defined, which is the minimum number of bits required to describe S_h. That is, $R_h(D)$ is the infimum of R_h such that there exist maps: $f_h : \mathcal{S}_h^n \rightarrow \left\{1, 2, \ldots, 2^{nR_h}\right\}$ and $g_h : \left\{1, 2, \ldots, 2^{nR_h}\right\} \times \mathcal{S}_v^n \rightarrow \hat{\mathcal{S}}_h^n$ such that

$$\lim_{n\to\infty} \sup \mathbb{E}[d(S_h^n, g_h(f_h(S_h^n), S_v^n))] \leq D, \tag{3.15}$$

where $d(\cdot)$ represents the distortion measure.

Based on the defined rate-distortion function, let sources (S_h, S_v, U) be $i.i.d. \sim p_{S_h,S_v}(s_h, s_v)p_{U|S_v}(u|s_v)$ and with $\mathcal{P} = \{P_j\}_1^M$. The distortion is measured by

$$d(S_h^n, \hat{S}_h^n) = \frac{1}{n}\sum_{i=1}^{n} d(s_{h,i}, \hat{s}_{h,i}). \tag{3.16}$$

When $R_v \geq I_r(S_v; U)$, the rate-distortion function $R_h^*(D)$ is given by

$$R_h^*(D) = \min_{p_{Z|S_h}(z|s_h)} \min_g (I_s(S_h; Z) - I_s(U; Z)). \tag{3.17}$$

Importantly, the minimization is over all functions $g = \mathcal{U} \times \mathcal{Z} \to \hat{\mathcal{S}}_h$ and conditional probability density function $p_{Z|S_h}(z|s_h)$, $|\mathcal{Z}| \leq |\mathcal{S}_h| + 1$, such that

$$\sum_{s_h\in\mathcal{S}_h, u\in\mathcal{U}, z\in\mathcal{Z}} p_{S_h,U}(s_h, u)p_{Z|S_h}(z|s_h)d(s_h, g(u, z)) \leq D. \tag{3.18}$$

The proof of this theorem can be found in [9].

Besides these two theorems mentioned above, several criteria can be given to verify the benefit of the cross-modal coding scheme in a quantitative way. The one criterion is cross-modal wealth $\mathcal{W}$, denoted as bits compressed by the CMC model relative to the baseline model, $\mathcal{W} = R_h' - R_h$. To be more specific, the expression of $\mathcal{W}$ is

$$\mathcal{W} = \begin{cases} I_r(S_h; U) - H_f(S_h|U), \ D = 0, \\ I_r(S_h; \hat{S}_h) - I_s(S_h; Z|U), \ D > 0, \end{cases} \tag{3.19}$$

where U is the decoded result of S_v, and Z is defined in Eq. (3.17). If S_h and S_v have strong semantic correlation, $I_r(S_h; U) > H_f(S_h|U)$ or $I_r(S_h; \hat{S}_h) > I_s(S_h; Z|U)$, it indicates that the constructed CMC model performs better than the baseline model. In contrast, if this is not the case, the baseline model proves to be superior. Moreover, the other criterion is the cross-modal compression ratio ρ, denoted as the ratio between R_h and R_h'. To be more specific, the expression of ρ is

$$\rho = \begin{cases} \dfrac{H_s(S_h|U)}{H_r(S_h)}, \ D = 0, \\ \dfrac{I_s(S_h; Z|U)}{I_r(S_h; \hat{S}_h)}, \ D > 0. \end{cases} \tag{3.20}$$

3.3.3 *Cross-Modal Codec Design*

Based on the theoretical results described above, a cross-modal codec can be designed. The aim is not only to leverage the benefit of the CMC model but also to make it compatible with current visual and haptic codecs. Specifically, the desired haptic signal can be assumed to be regenerated from reference samples through channel coding techniques. Consequently, the LDPC codes [12] can be employed here. These reference samples are deduced from the associated visual frames by leveraging the correlation between haptic and visual semantics with the aid of deep learning [13]. By approximating the reference samples to be sufficiently close to the actual samples, a reduction in the sampling rate required for the desired haptic signal may be anticipated. The framework of the designed cross-modal codec is depicted in Fig. 3.8. Notably, the two theorems in Sect. 3.3.2 delineate the interplay between the rate of transmission and the degree of distortion for a pair of visual-haptic sources across all levels of distortion (i.e., $D \geq 0$). This forms the theoretical foundation for the rate control mechanism of this designed cross-modal codec. This rate-distortion correlation will inform the encoder on the appropriate bit allocation for both haptic and visual streams to satisfy the decoder's distortion specifications. For the visual frames, HEVC encoders and decoders are utilized for independently processing them, which can be referred to in the HEVC standard. Next, we outline the sequence of operations within this framework. The focus of the design is primarily on the encoding of haptic signals and how visual modality can assist in haptic decoding.

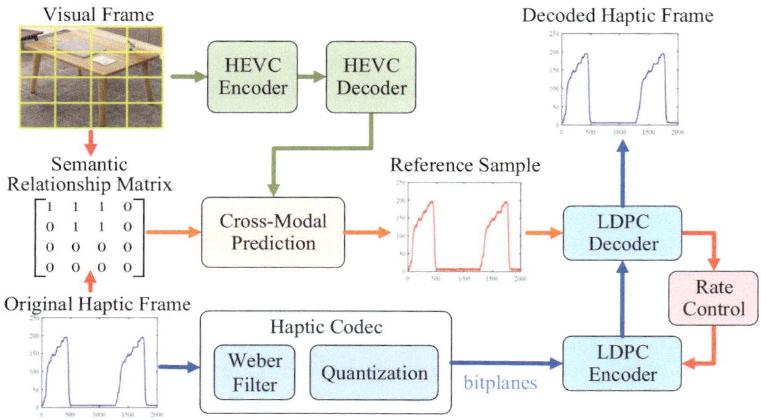

Fig. 3.8 The framework of the proposed cross-modal codec

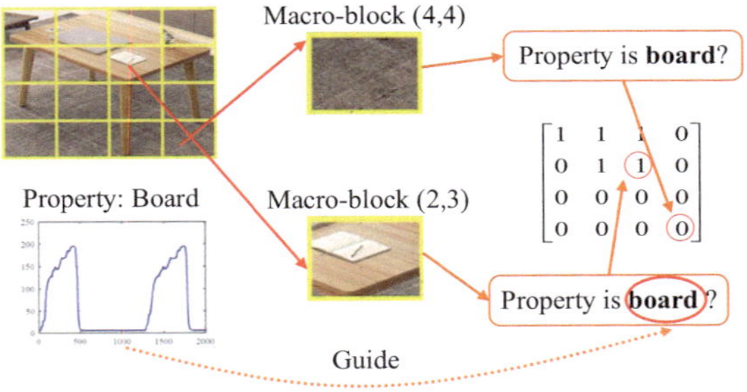

Fig. 3.9 The illustration of how to get the semantic relation matrix

Encoding

The haptic and visual inputs are encoded frame by frame. We assume that each haptic frame contains n_h samples, resulting in a duration of $T_h = n_h / f_h$, with f_h being the sampling frequency of the haptic signal. Likewise, the duration of each visual frame is $T_v = 1/f_v$, where f_v is the related visual frame rate. Consequently, one visual frame corresponds to $\lfloor \frac{T_v}{T_h} \rfloor$ haptic frames.

To enable the decoder to predict reference samples, a semantic relationship matrix between each pair of haptic and visual frames must be conveyed within the bitstream. As depicted in Fig. 3.9, elements of this matrix indicate whether each macroblock aligns with the "stone" property in the current visual frame, that is, the membership of the macroblock in possessing the "stone" attribute. It is noted that these elements can be manually annotated or derived by a carefully designed algorithm. Given the semantic relationship matrix's minimal data volume and infrequent changes, it is placed at the beginning of each or several groups of visual frames.

When it comes to haptic coding, the Weber filter represents a canonical compression method for haptic signals. Signals are perceivable and require transmission only if the relative difference between consecutive stimuli surpasses the just-noticeable difference. The Weber filter's output signals are subject to uniform quantization. Finally, the quantization symbols of the same bit value are aggregated to form a bitplane, starting from the most significant bit to the least significant bit.

When considering the details of LDPC, quantization symbols across N bitplanes are encoded plane by plane. The most significant bitplane is first coded by the LDPC encoder, followed by the less significant bitplanes in sequence. For each bitplane encoding, it is multiplied by a crafted LDPC check matrix to yield the corresponding syndrome. For an (n, k) LDPC code, if the length of each bitplane is n, the length of its parity bits will be $n - k$. Ultimately, these parity bits are transmitted to the decoder.

Decoding

The visual frames are decoded by HEVC. After combining the decoded visual frames and the semantic relationship matrix between two modalities, the reference frame of haptic modality can be predicted, shown as the cross-modal prediction module in

Fig. 3.8. Specifically, the received semantic relationship matrix is utilized to identify the haptic-located block that shares the same visual-haptic characteristics within the current visual frame. Following this, methods such as generative adversarial network (GAN), diffusion model, etc., are utilized to predict the reference samples based on the visual pixels within the identified haptic-located block. In this context, the effectiveness of the prediction method can be assessed by calculating the difference (e.g., root mean square error) between the actual haptic samples and the predicted reference samples.

After obtaining the received parity bits and the quantization of the reference samples, LDPC decoding is performed. Similar to LDPC encoding, decoding starts with the most significant bitplane and then proceeds to the less significant ones. The discrepancies between the reference samples and the original haptic samples are corrected using parity bits, which allows the recovery of the original haptic frame from the reference samples. The decoding process for each bitplane follows a similar method to that described in [12]. Importantly, a rate control scheme is implemented here, supported by a feedback channel. To be more specific, the decoder firstly estimates the required rate according to the quality of the derived reference samples and delivers it back to the encoder. Then, the encoder sends the parity bits according to the received rate demands. The decoder adopts these additional parity bits for decoding. If the decoding fails, the decoder can request additional parity bits until the bitplane is successfully decoded. Once LDPC decoding is finished, all bitplanes are merged to obtain the haptic frame.

3.3.4 Experimental Result and Analysis

In this section, the validation of the proposed theoretical theorems about the cross-modal coding model in Sect. 3.3.2 is employed for binary sources. Then, the performance of the proposed cross-modal codec in Sect. 3.3.3 is evaluated in a practical teleoperation system.

Theoretical validation

In the simulated environment, two binary sources about haptic modality S_h and visual modality S_v are respectively considered. The collection of properties for both haptic and visual is identical, represented as $\mathcal{P}h = \mathcal{P}v = \{P_j\}_{j=1}^{10}$. Two probabilities are established. The probability that $S_{h,i}$ and $S_{v,i}$ exhibit different symbols is $Pr(s_{h,i} \neq s_{v,i}) = p_1$. The probability that they possess distinct semantic properties is $Pr(P_{h,i} \neq P_{v,i}) = p_2$. An event E is also defined, where $E = 1$ if $S_{h,i}$ and $S_{v,i}$ are not identical in either their symbols or properties, and $E = 0$ otherwise. Consequently, the probability of $E = 1$ is $p = 1 - (1 - p_1)(1 - p_2)$, and the probability of $E = 0$ is $1 - p$. For theoretical validation, the bit rate for each visual source is ensured to meet $R_v \geq H_r(S_v)$. Under this setting, the minimum bit rate R_v required for each haptic source corresponds to $H_s(S_h|U) = H_s(S_h|S_v)$. Based on the defini-

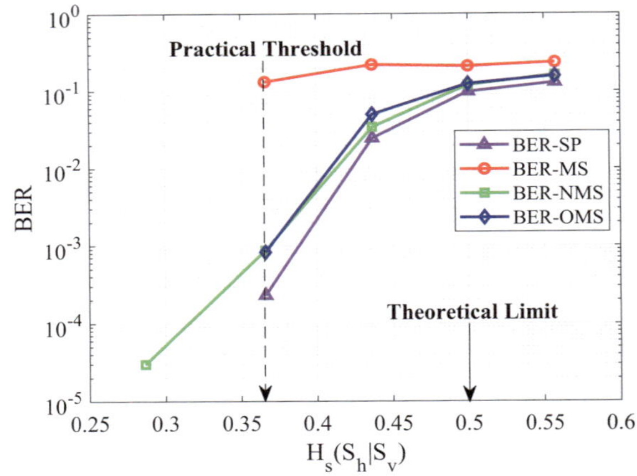

Fig. 3.10 Theoretical validation about coding efficiency

tions of E and $H_s(S_h|S_v)$, the minimum bit rate R_v can be determined by

$$H_s(S_h|S_v) = H_s(E) = \underbrace{p \log \frac{1}{p}}_{H_s(E=1)} + \underbrace{(1-p) \log \frac{1}{1-p}}_{H_s(E=0)}. \tag{3.21}$$

Moreover, in order to evaluate the LDPC code-based haptic compression performance, several representative decoding algorithms such as sum-product (SP), min-sum (MS), normalized min-sum (NMS), and offset min-sum (OMS) are utilized.

We initially assess the coding efficiency. The theoretical limit presented in Eq. (3.13) as a reference point, and the comparative outcomes are depicted in Fig. 3.10. It is evident that the actual cross-modal codec performance closely approximates the theoretical limit. At a cross-modal compression ratio of $\rho = 0.5$, the threshold for the proposed cross-modal codec is $H_s(S_h|S_v) = 0.366$ bits, ensuring that the bit error rate (BER) for $S_{h,i}$ is nearly negligible. To highlight the improvements achieved by cross-modal coding, three rate-distortion curves based on BER and bit count are also provided when $H_s(S_h|S_v) = 0.5$ bits. As illustrated in Fig. 3.11, the CMC model exhibits significant enhancement over the baseline model. Specifically, the relative benefit of cross-modal coding is capped at 36.6%, and in practical scenarios, it does not exceed 50%. These findings indicate that the proposed cross-modal codec is close to achieving the optimal compression efficiency.

Performance evaluation and analysis

The proposed cross-modal codec in Sect. 3.3.3 is implemented on a self-developed practical teleoperation platform shown in Fig. 3.12. This platform is developed based on the cross-modal communication architecture proposed in Sect. 2.3. The related dataset generated by this platform will be introduced in Sect. 6.1. For the experiment,

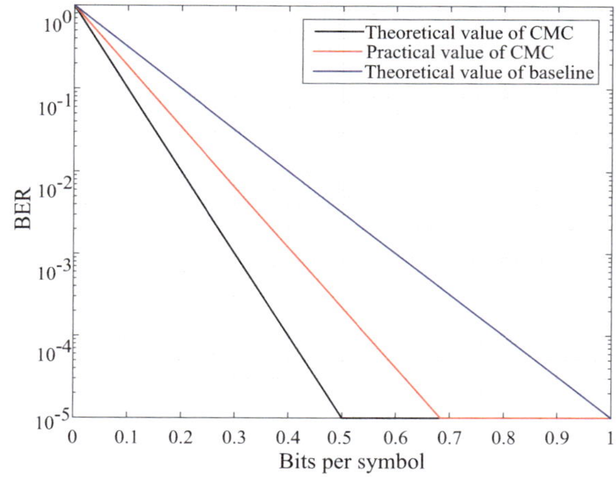

Fig. 3.11 Theoretical validation about rate distortion

a haptic frame consists of 80 haptic samples. One thousand five hundred pairs of haptic and visual frames are utilized for testing purposes. The distortion tolerance for haptic signals is set to $D = 0$. The performance of the cross-modal prediction method is assessed using the root mean square error (RMSE) between the actual haptic sample $s_{h,i}$ and the reference haptic sample $s_{h,i}'$. The minimum number of bits $H_s(S_h|U)$ required for each haptic sample is determined by $RMSE$, the size of the semantic relationship matrix C, and the number of haptic samples in one transmission N_h. The formulation of $H_s(S_h|U)$ is

$$\begin{aligned} H_s(S_h|U) &= H_r(S_h|S_h') + H_f(S_h|U) \\ &= K_1 + K_2 \log_2\left(\frac{K_3}{RMSE^2}\right) + \frac{C}{N_h}, \end{aligned} \tag{3.22}$$

where K_1, K_2 and K_3 are parameters. The haptic source is assumed to obey a Gaussian distribution. If $C = 16$ and $N_h = 800$, it means each visual frame is divided into 4×4 macroblocks, and a $16/800$ bits message records the semantic relationship between visual macroblocks and each haptic sample.

Figure 3.13 demonstrates the effectiveness of cross-modal prediction by showing the actual haptic samples with their corresponding reference samples. As depicted in Fig. 3.13a, a Weber filter is initially applied to sift through 10 haptic frames (each frame contains 80 samples), totaling 800 samples, and 560 perceptible haptic samples as selected for transmission. Figure 3.13b presents the associated reference samples. With the haptic and reference frames in hand, Fig. 3.14 displays the bitplanes of the haptic samples along with their LDPC encoding. Each haptic sample value is quantized to 4 bits, resulting in a quantization error of 0.125, which allows for the extraction of 4 bitplanes. Each bitplane is then multiplied by a rate-adaptive check matrix to generate parity bits. The higher bitplane of the haptic frame and the reference frame exhibit a relatively low bit error, leading to a smaller number of parity

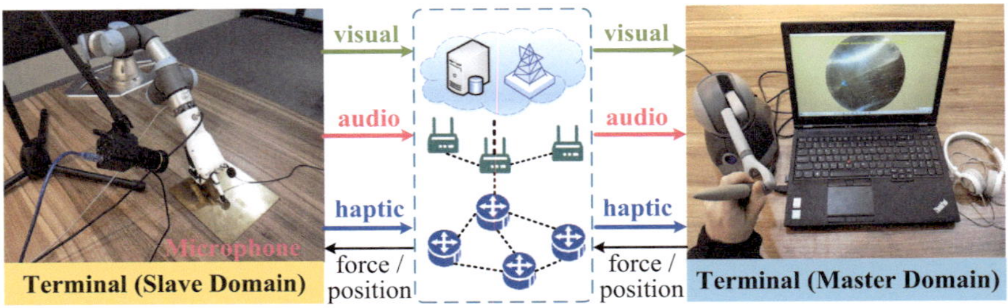

Fig. 3.12 The illustration of the constructed practical teleoperation system

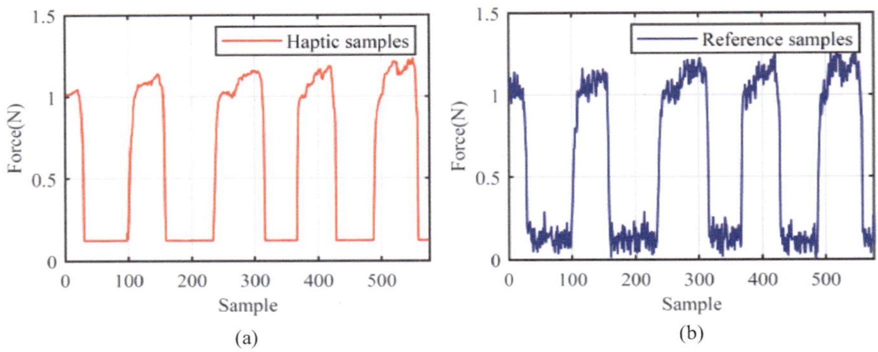

Fig. 3.13 The performance of the cross-modal prediction: **a** The haptic samples passing through the Weber filter. **b** The reference samples generated from the decoded visual frames

bits. Conversely, the lower bitplane has a higher number of parity bits. Overall, the transmission of samples has been substantially reduced.

Then, we examine the performance of the proposed cross-modal codec in Sect. 3.3.3 for practical haptic and visual signals. Here, different cross-modal prediction RMSEs can be obtained by using visual frames with different semantic relationship matrices to predict reference samples. Through a lot of experiments under a various of RMSE, the parameter values of K_1, K_2 and K_3 are determined in Eq. (3.22), and the $H_s(S_h|U)$-$RMSE$ curve is fitted in Fig. 3.15a. Moreover, considering the two criteria $\mathcal{W}$ and ρ in Eqs. (3.19) and (3.20), we plot the $\mathcal{W}$-RMSE curve and the ρ-$RMSE$ curve in Fig. 3.15b–c based on $H_s(S_h|S_v)$-$RMSE$ curve. Given these results, we find two important conclusions. On one hand, $H_s(S_h|U)$ increases as $RMSE$ increases. On the other hand, the performance of the CMC model deteriorates as $RMSE$ increases. Moreover, we adjust the bitrate of the CMC model by setting the number of bitplanes, while the bitrate of the baseline model is controlled by the quantization parameter. The rate-distortion performance of haptic samples is measured by the $PSNR$-bitrate curve. As shown in Fig. 3.15d, the CMC model has a considerable improvement compared to the baseline model. Finally, we collect haptic and visual signals for transmission in real time. Figure 3.16 shows the col-

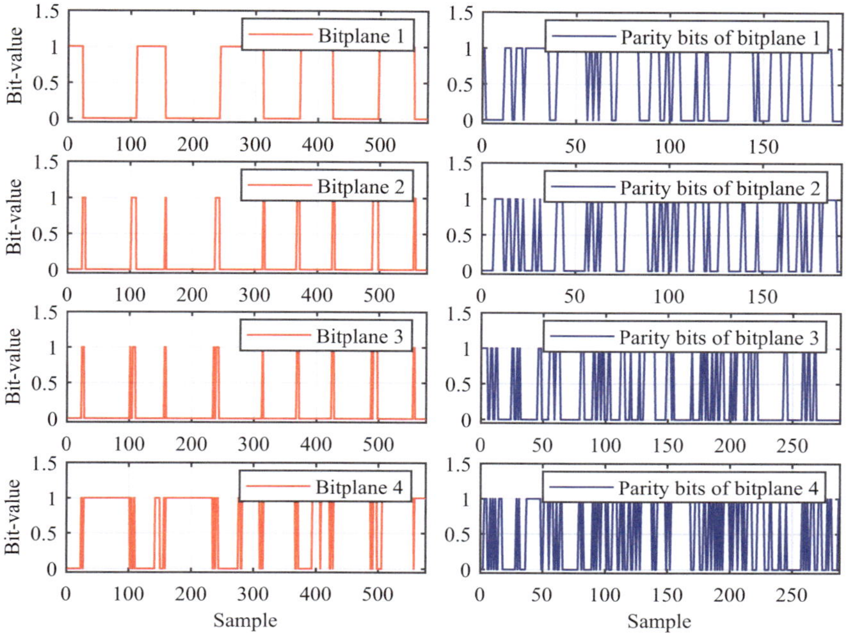

Fig. 3.14 The illustration of bitplanes and their LDPC encoding

lected pressure signals and the performance comparison of multi-modal streams by the baseline model and CMC. Given the results, it can be seen that the impact on the visual stream can be alleviated by the CMC model.

3.4 Visual-Haptic Mutual Aided Coding

From the above sections, by using potential correlations among modalities, redundancies in the multi-modal source signals can be reduced, and one modality signal can be compressed with the aid of the other modality signal. However, existing schemes are not yet mature. On the one hand, the optimal rate under given delay and reliability constraints is still unknown. On the other hand, under the optimum rate, how to design an available cross-modal coding (CMC) scheme that realizes mutual compression among modalities. Therefore, this section seeks to tackle the challenge of efficient and dependable source coding for multi-modal sources by the optimum rate of a general CMC and its practical implementation [14]. The expression of the achievable rate region for the general CMC is first derived. Then, based on this region and considering delay and reliability constraints, the optimum rate can be obtained. Finally, a comprehensive realization of a visual-haptic mutual-aided

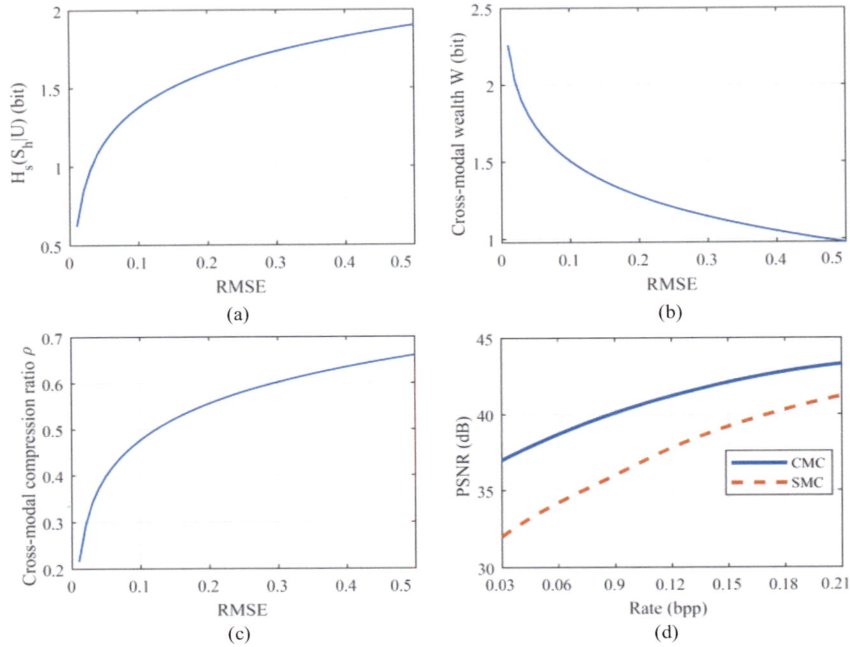

Fig. 3.15 Coding performance: **a** Minimum number of bits for each haptic sample; **b** Cross-modal wealth $\mathcal{W}$; **c** Cross-modal compression ratio ρ; **d** Rate distortion curve

codec is designed by fully harnessing the capabilities of channel coding and artificial intelligence-based semantic analysis.

3.4.1 General Cross-Modal Coding Model and Entropy Density

The constructed general CMC model is depicted in Fig. 3.17. Similar to Sect. 3.3, a pair of semantically correlated visual and haptic signals is taken as an example. The encoders observe a pair of discrete memoryless visual and haptic sources, denoted as (S_v, P_v) and (S_h, P_h). They are with a specified joint probability distribution P_{S_v,P_v,S_h,P_h} over the combined alphabet space $\mathcal{S}_v \times \mathcal{P}_v \times \mathcal{S}_h \times \mathcal{P}_h$. It is noted that S_v^n, P_v^n, S_h^n, P_h^n respectively represent visual symbol, the related visual semantics, haptic symbol, and the related haptic semantics, where n represents the sequence's block length. This general CMC model employs a standard block coding framework [15], which compresses these n-length sequences of sources into a rate-limited codeword. The correlation between visual and haptic modalities can be described as the relationship between P_v and P_h, as only signals with consistent semantic types exhibit

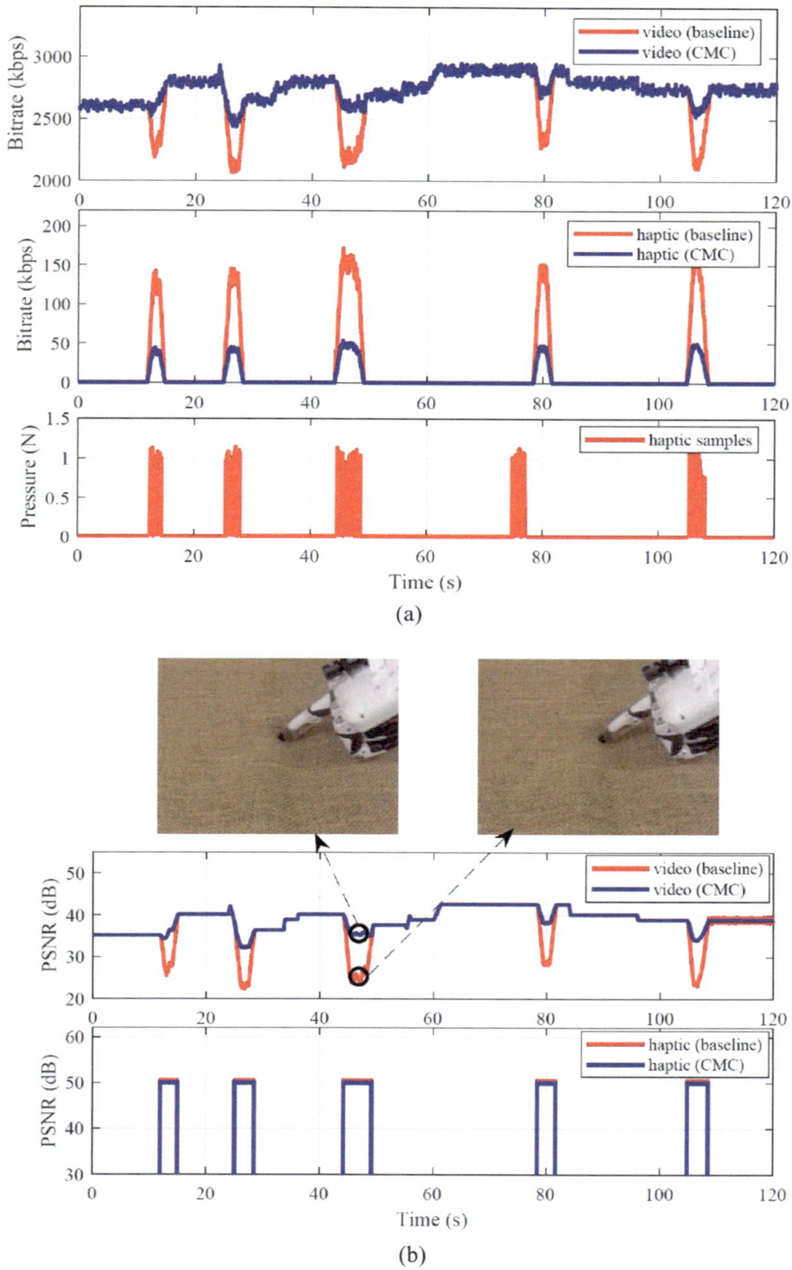

Fig. 3.16 Performance comparison: **a** Rate; **b** Distortion

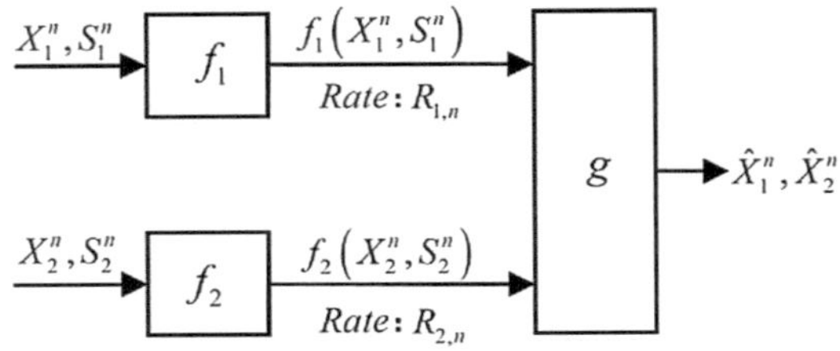

Fig. 3.17 Illustration of the general CMC [14]

semantic redundancy that requires further compression. Thus, both the symbols and their semantic types must be encoded. This allows the decoder to utilize the semantic correlation fully to recover the original signals. Here, the recovery of semantics P_v and P_h is assumed to be lossless. Consequently, the overall probability of error can be represented as $P_e = \mathbb{P}[(\hat{S}_v^n, \hat{S}_h^n) \neq (S_v^n, S_h^n)]$, where $\hat{S}_v^n, \hat{S}_h^n$ are the recovered sequences of symbols. Based on this, the general CMC model can be defined in the following.

General CMC model: An $(n, 2^{nR_{v,n}}, 2^{nR_{h,n}}, \varepsilon)$-general CMC model for the visual-haptic modality sources (S_v, P_v, S_h, P_h) consists of two encoding functions $f_v : \mathcal{S}_v^n \times \mathcal{P}_v^n \to \{1, 2, \ldots, 2^{nR_{v,n}}\}$, $f_h : \mathcal{S}_h^n \times \mathcal{P}_h^n \to \{1, 2, \ldots, 2^{nR_{h,n}}\}$, and a decoding function $g : \{1, 2, \ldots, 2^{nR_{v,n}}\} \times \{1, 2, \ldots, 2^{nR_{h,n}}\} \to \mathcal{S}_v^n \times \mathcal{S}_h^n$. The aim is to make the error probability P_e no greater than ε, where ε is the target error probability. $R_{v,n}$ and $R_{v,n}$ denote the source coding rates required to describe (S_v, P_v) and $S_h, P_h)$, respectively. Moreover, the related source coding rate vector $\mathbf{R} \in \mathbb{R}^3$ can be denoted as

$$\mathbf{R} = \begin{bmatrix} R_{v,n} \\ R_{h,n} \\ R_{v,n} + R_{h,n} \end{bmatrix}. \tag{3.23}$$

Moreover, to conveniently verify the cross-modal coding model, the baseline model should also be given in the following.

Baseline model: An $(n, 2^{nR'_{v,n}}, 2^{nR'_{h,n}})$ code for sources (S_v, S_h) comprises two encoding functions $f'_v : \mathcal{S}_v^n \to \{1, 2, \ldots, 2^{nR'_{v,n}}\}$, $f'_h : \mathcal{S}_h^n \to \{1, 2, \ldots, 2^{nR'_{h,n}}\}$, and two decoding functions $g_v : \{1, 2, \ldots, 2^{nR'_{v,n}}\} \to \mathcal{S}_v^n$, $g_h : \{1, 2, \ldots, 2^{nR'_{h,n}}\} \to \mathcal{S}_h^n$, where $R'_{v,n}$ and $R'_{h,n}$ denote the rates required to describe S_v and S_h, respectively. As this baseline model does not consider the semantic correlation between visual and haptic modality sources, $R_{v,n} \geq H(S_v)$ and $R_{h,n} \geq H(S_h)$ should be satisfied.

Next, the *entropy density vector* is constructed as

$$\mathbf{h}(S_v, P_v, S_h, P_h) = \begin{bmatrix} -\log P_{S_v, P_v|S_h, P_h}(S_v, P_v|S_h, P_h) \\ -\log P_{S_h, P_h|S_v, P_v}(S_h, P_h|S_v, P_v) \\ -\log P_{S_v, P_v, S_h, P_h}(S_v, P_v, S_h, P_h) \end{bmatrix}. \tag{3.24}$$

Based on this, the mathematical expectation of the entropy density vector is

$$\mathbf{H}(S_v, P_v, S_h, P_h) = \mathbb{E}[\mathbf{h}(S_v, P_v, S_h, P_h)] = \begin{bmatrix} H(S_v, P_v|S_h, P_h) \\ H(S_h, P_h|S_v, P_v) \\ H(S_v, P_v, S_h, P_h) \end{bmatrix}. \tag{3.25}$$

Subsequently, the covariance of the entropy density vector $\mathbf{V}(X_1, S_1, X_2, S_2)$ is

$$\mathbf{V}(S_v, P_v, S_h, P_h) = Cov[\mathbf{h}(S_v, P_v, S_h, P_h)]. \tag{3.26}$$

For expression convenience, $\mathbf{h}(S_v, P_v, S_h, P_h)$, $\mathbf{H}(S_v, P_v, S_h, P_h)$ and $\mathbf{V}(S_v, P_v, S_h, P_h)$ are respectively abbreviated as $\mathbf{h}$, $\mathbf{H}$ and $\mathbf{V}$. Finally, κ is defined as the maximum of the 2-norm of the Hessians of each element in $\mathbf{H}$,

$$\kappa \triangleq \max_{1\le i\le 3} \left\| \nabla^2_{P_{S_v,P_v,S_h,P_h}} \mathbf{H}_{(i,1)} \right\|_2. \tag{3.27}$$

3.4.2 Coding Theory Under Achievable Rate Region

Based on the constructed general CMC model and the defined entropy density, an achievable rate region is required by considering the latency and reliability constraints. This region is needed to satisfy the desired fidelity and to sustain the target error probability at a fixed block length. After obtaining the infimum of this region, the minimum source coding rate demanded for reliable visual and haptic streaming can be determined.

Achievable rate region: A rate pair $(R_{v,n}, R_{h,n})$ is said to be (n, ε)-achievable if there exists an $(n, 2^{nR_{v,n}}, 2^{nR_{h,n}}, \varepsilon)$-general CMC code with $P_e \le \varepsilon$. The (n, ε)-achievable rate region $\mathcal{R}^*_{\text{generalCMC}}(n, \varepsilon) \subset \mathbb{R}^2$ is the closure of the set of (n, ε)-achievable rates.

Then, the set $\varphi(\mathbf{V}, \varepsilon) \triangleq \{\mathbf{w} \in \mathbb{R}^3 : \mathbb{P}[\mathbf{W} \le \mathbf{w}] \ge 1 - \varepsilon\}$ is defined. This set is concerned as the cumulative distribution function of a multivariate Gaussian with zero mean and covariance $\mathbf{V}$. Based on this set, the inner bound region, outer bound region of (n, ε)-achievable rate, and the achievable rate region can be derived in the following.

Inner bound region: The inner bound region $\mathcal{R}_{\text{in}}(n, \varepsilon) \subset \mathbb{R}^2$ is the set of all the rates $(R_{1,n}, R_{2,n})$ satisfying

$$\mathbf{R} \in \mathbf{H} + \frac{1}{\sqrt{n}}\varphi(\mathbf{V}, \varepsilon) + \frac{\alpha \log n}{n}\mathbf{I}, \tag{3.28}$$

where $\alpha = \Pi_{i=1,2} |\mathcal{S}_i| \, \Pi_{j=1,2} \left|\mathcal{P}_j\right| + \kappa + 3/2$.

Outer bound region: The outer bound region $\mathcal{R}_{\text{out}}(n, \varepsilon) \subset \mathbb{R}^2$ is the set of all the rates fulfilling

$$\mathbf{R} \in \mathbf{H} + \frac{1}{\sqrt{n}} \varphi (\mathbf{V}, \varepsilon) - \frac{\log n}{n} \mathbf{I}. \tag{3.29}$$

Achievable rate region: For (S_v, P_v, S_h, P_h) drawn i.i.d $\sim P_{S_v, P_v, S_h, P_h}$, the (n, ε)-achievable rate region $\mathcal{R}^*_{\text{generalCMC}}(n, \varepsilon)$ obeys

$$\mathcal{R}_{\text{in}} (n, \varepsilon) \subset \mathcal{R}^*_{\text{generalCMC}} (n, \varepsilon) \subset \mathcal{R}_{\text{out}} (n, \varepsilon) . \tag{3.30}$$

Subsequently, the optimum rate pair $(R^*_{v,n}, R^*_{h,n})$ is the infimum of all the (n, ε)-achievable rates $(R_{v,n}, R_{h,n})$ in $\mathcal{R}^*_{\text{generalCMC}}(n, \varepsilon)$. The relationship between this optimum rate pair and the certain boundary points of a general CMC rate region can be described by a rate dispersion function [9]. To be more specific, this can be approached from the boundary of $\mathcal{R}^*_{\text{generalCMC}}(n, \varepsilon)$ at a specific angle θ. For an $(n, 2^{nR_v}, 2^{nR_h}, \varepsilon)$-general CMC code, the rate dispersion function F^* is the infimum of all functions F satisfying

$$\begin{aligned} \limsup_{n\to\infty} \sqrt{n} \left(R_{v,n} - R^*_v \right) &\leq \sqrt{F}(\cos\theta) Q^{-1} (\varepsilon) , \\ \limsup_{n\to\infty} \sqrt{n} \left(R_{h,n} - R^*_h \right) &\leq \sqrt{F}(\sin\theta) Q^{-1} (\varepsilon) , \end{aligned} \tag{3.31}$$

where $R^*_v = H(S_v, P_v | S_h, P_h)$, $R^*_h = H(S_h, P_h | S_v, P_v)$ and $\mathcal{R}^*_{\text{generalCMC}} = \{\mathbf{R} : \mathbf{R} \in \mathbf{H}\}$ [9].

From the above theorem, the optimum rates $(R^*_{1,n}, R^*_{2,n})$ required to hold ε can fulfill

$$\begin{aligned} R^*_{v,n} &\approx R^*_v + \sqrt{\frac{F^*}{n}} (\cos\theta) \, Q^{-1} (\varepsilon) , \\ R^*_{h,n} &\approx R^*_h + \sqrt{\frac{F^*}{n}} (\sin\theta) \, Q^{-1} (\varepsilon) . \end{aligned} \tag{3.32}$$

Finally, an approximation for the optimum rate pair $(R^*_{v,n}, R^*_{h,n})$ can be obtained.

Optimum rate: For (S_v, P_v, S_h, P_h), the optimum rates can be approximated as

$$\begin{aligned} R^*_{v,n} &\approx H (S_v, P_v | S_h, P_h) + \sqrt{\frac{\mathbf{V}_{(1,1)}}{n}} Q^{-1} (\varepsilon) , \\ R^*_{h,n} &\approx H (S_h, P_h | S_v, P_v) + \sqrt{\frac{\mathbf{V}_{(2,2)}}{n}} Q^{-1} (\varepsilon) , \\ R^*_{v,n} + R^*_{h,n} &\approx H (S_v, P_v, S_h, P_h) + \sqrt{\frac{\mathbf{V}_{(3,3)}}{n}} Q^{-1} (\varepsilon) . \end{aligned} \tag{3.33}$$

The proof of the above theorems can be found in [14].

3.4.3 General Cross-Modal Codec Design

Based on the theoretical results described above, a general cross-modal codec can be designed by fully utilizing AI semantic analysis and channel coding. The approach involves employing channel coding to condense the target modality signal into parity bits that facilitate the reconstruction of the original signal from the cross-modal prediction signal. In the signal compression phase, the encoder subjects the target modality signal to channel encoding, yielding parity bits as the output of the source encoding process. In the signal reconstruction phase, leveraging an AI-driven cross-modal prediction network, a predicted signal with the target modality is generated from the correlated signals with the other modalities. Finally, the decoder utilizes the received parity bits to refine the predicted signal, thereby obtaining the original signal. If the predicted signal closely resembles the original signal, a higher compression ratio can be achieved with parity bits at a reduced coding rate. Importantly, this generic cross-modal codec that realizes visual-haptic mutual coding can be divided into two main components, the cross-modal visual codec and the cross-modal haptic codec.

Cross-modal visual codec (CMVC)

The core concept of CMVC involves utilizing the accompanying haptic signal to compress visual signals that share potential correlations or semantics. It can be further divided into two stages: signal compression during the encoding procedure (shown in Fig. 3.18) and signal reconstruction during the decoding procedure (shown in Fig. 3.19).

Encoding: Here, the input visual and haptic signals, along with the semantic types deduced from these two modalities, are encoded in isolation. For the visual signal, each frame is initially processed by a classifier. This classifier is to detect whether a visual macroblock is consistent with the semantic type of the associated haptic signal. The visual macroblocks inconsistent with the haptic signal are encoded utilizing intraframe compression techniques such as HEVC [7], VVC [16]. On the contrary, those visual macroblocks that are semantically linked to or consistent with the haptic signal are compressed using the polar code method [17]. Specifically, the process begins with uniform quantization of the visual macroblocks, followed by the extraction of all quantization symbol bits to create bitplanes. The bitplanes of the visual macroblocks are then encoded layer by layer. Each bitplane is encoded by combining it with the generator matrix of a structured systematic polar code, resulting in the output of only parity bits. These parity bits are then transmitted to the decoder. If each bitplane has a length of n, compression is achieved when the number of parity bits is less than or equal to n using a (m, n) polar code, with $m < 2n$. It's important to note that the value of m is determined by rate control, as per Eq. (3.33).

Regarding the haptic signal, the perceptual deadband method is chosen for haptic source coding, drawing on principles from Weber's law. Finally, for the semantic types, to produce the prediction signal at the decoder, semantic residual encoding is

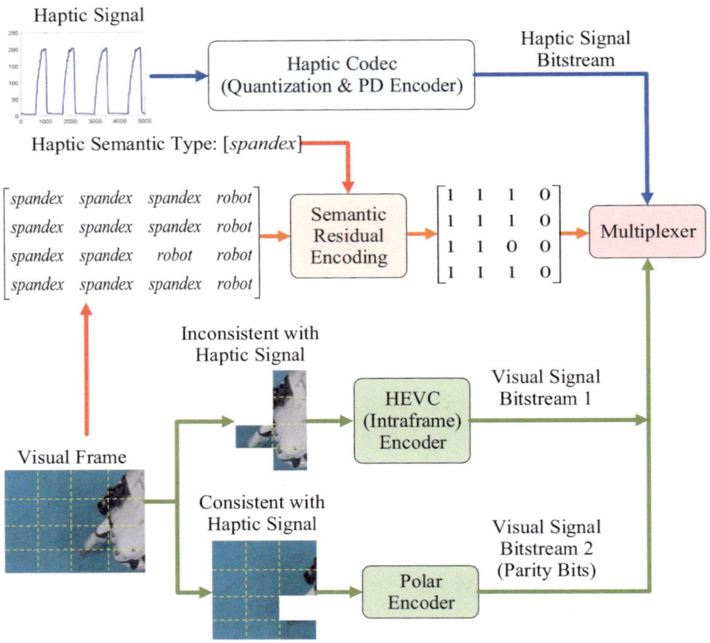

Fig. 3.18 Signal compression during encoding procedure of the CMVC

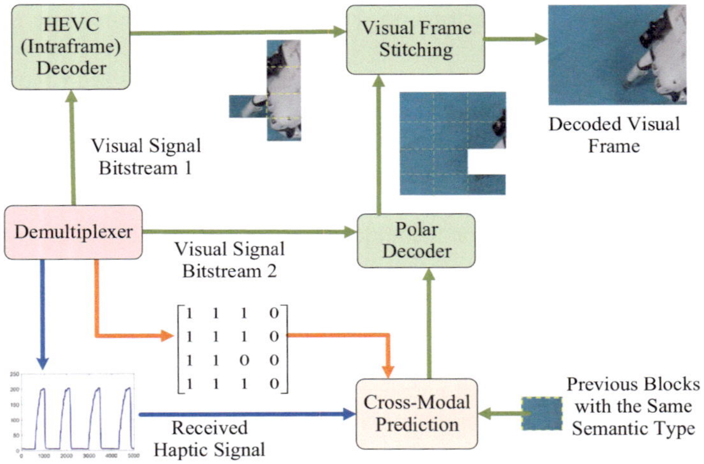

Fig. 3.19 Signal reconstruction during decoding procedure of the CMVC

performed to ensure the availability of semantic relevance between visual and haptic modalities to the decoder. The main idea is to encode the discrepancy between the semantic type of each visual macroblock and that of the haptic signal. As illustrated in Fig. 3.18, semantic relevance is represented by a message matrix, the dimensions of which are based on the macroblocks within a visual frame. Each matrix element's value indicates whether the semantic type of each visual macroblock aligns with the semantic type "spandex" of the coexisting haptic signal. A value of 1 signifies alignment, while 0 indicates otherwise. Significantly, semantic relevance can be conveyed every few frames due to the minimal change in semantic types over brief periods.

Decoding: In the decoding procedure, the haptic signal, semantic correlations, and visual macroblocks inconsistent with the haptic signals are readily retrieved from their respective bitstreams. In particular, in Fig. 3.19, visual macroblocks inconsistent with the haptic signals are retrieved directly from visual signal bitstream 1 through intraframe decoding. In contrast, visual macroblocks consistent with the haptic signal are extracted from the visual signal bitstream 2, which contains parity bits, utilizing AI semantic analysis and polar code decoding techniques. The objective of AI semantic analysis is to produce a predicted visual signal leveraging semantic correlations. To be more specific, the semantic type of the haptic signal is deduced from the received signal, and then the semantically related visual macroblocks are identified using the semantic relevance message matrix. For instance, in Fig. 3.19, the $1st$ macroblock is linked to the haptic signal with a semantic type of "spandex", indicated by a value of 1 in the matrix's $1st$ element. The predicted visual signal for these macroblocks is generated from the previously decoded visual macroblocks and the haptic signal sharing the same semantic type, employing the cross-modal prediction network, such as that in Sect. 5.2 [18].

After obtaining the predicted visual signal, it is treated as a noisy rendition of the original macroblocks consistent with the haptic signal. In other words, upon receiving the parity bits, the original visual signal's bitplanes are decoded sequentially in the polar code decompression process. The decoder firstly concatenates each bitplane of the prediction signal with its corresponding parity bits. Then, by employing an algorithm such as successive cancellation list decoding, the decoder processes the concatenated bits to achieve a soft estimate for each bitplane of the original visual signal. Ultimately, the decoder multiplies the estimation by the check matrix to confirm its precision. If the error probability falls below a specified threshold, the decoder requests additional parity bits from the encoder. Finally, by assembling all the visual macroblocks, the original visual signal can be reconstructed.

Cross-modal haptic codec (CMHC)

In Figs. 3.20 and 3.21, the signal compression during the encoding procedure and signal reconstruction during the decoding procedure of CMHC mirror those of CMVC, albeit with certain modifications tailored to the nuances of haptic modality.

Encoding: Visual macroblocks are all subjected to intraframe encoding. The haptic signal undergoes polar code-based compression. To be more specific, the haptic signal

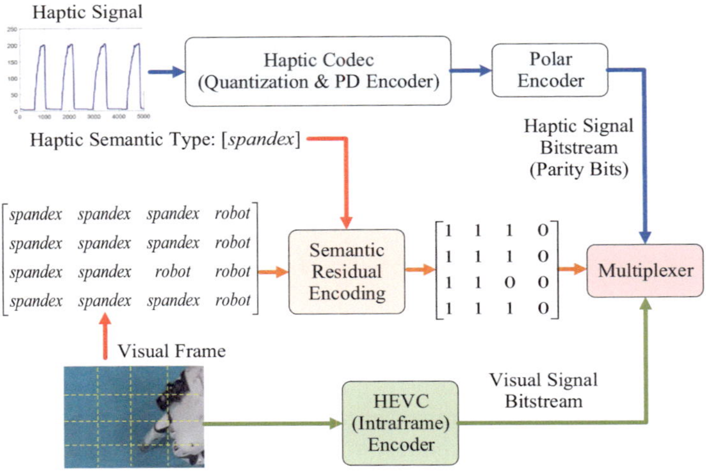

Fig. 3.20 Signal compression during encoding procedure of the CMHC

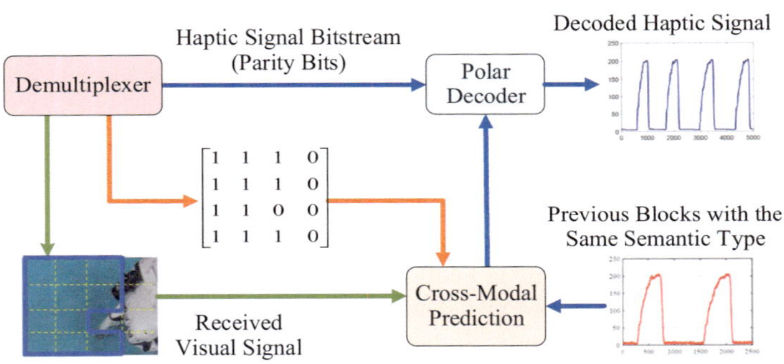

Fig. 3.21 Signal reconstruction during decoding procedure of the CMHC

initially traverses a Weber filter to isolate samples that exceed the just-noticeable difference from their preceding counterparts. Subsequently, the Weber filter's output samples are quantified, creating bitplanes. Ultimately, the bitplanes of the haptic quantization symbols are converted into parity bits through polar code encoding.

Decoding: Visual signal is straightforwardly reconstructed from its bitstream. Haptic signal necessitates reconstruction via AI-driven semantic analysis and decompression of polar codes. Regarding AI semantic analysis, the semantic type is deduced from the incoming visual signal, which then facilitates the determination of the haptic semantic type through semantic correlation. For example, the semantic correlation value being 1 indicates that the 1*st* macroblock is semantically linked to the haptic signal, inferring a haptic semantic type of "spandex", shown in Fig. 3.21. Thereafter, the predicted haptic signal is crafted from previously decoded haptic signals sharing the identical semantic type. Ultimately, the original haptic signal is reconstructed from this predicted signal, with parity bits adjusting the predicted signal samples that deviate from the original haptic signal.

3.4.4 Experimental Result and Analysis

The theoretical results proposed in Sect. 3.4.2 are firstly verified in a case concerning the binary symbol and semantic sequences. Then, the performance of the proposed general cross-modal codec in Sect. 3.4.3 is evaluated in a practical teleoperation system.

Theoretical validation

In the simulated environment, two i.i.d. equiprobable binary sources about visual modality (S_v^n, P_v^n) and haptic modality (S_h^n, P_h^n) are respectively considered, $\mathbb{P}\left[S_{v,i} \neq S_{h,i} \text{ or } P_{v,i} \neq P_{2,i}\right] = p \leq 0.5$. The rate of haptic source variable $(S_{h,i}, P_{h,i})$ is assumed to fulfill $R_h \geq H(S_h, P_h)$. Under this premise, $R_{v,n}^*$ about visual source can be computed by Eq. (3.33), obtaining

$$R_{v,n}^* = \mathbb{E}\left[-\log P_{S_v,P_v|S_h,P_h}\left(S_v, P_v|S_h, P_h\right)\right] \tag{3.34}$$

$$+\sqrt{\frac{\mathrm{Var}\left[-\log P_{S_v,P_v|S_h,P_h}\left(S_v, P_v|S_h, P_h\right)\right]}{n}}Q^{-1}\left(0\right), \tag{3.35}$$

where $-\log P_{S_v,P_v|S_h,P_h}\left(S_v, P_v|S_h, P_h\right) = -\log_2 p$ with probability p, and it equals to $-\log_2(1-p)$ with probability $1-p$. For verification, $R_{v,n}^*$ is obtained by using a rate 2/3 polar code and a decoding algorithm. Moreover, $H\left(S_v, P_v|S_h, P_v\right)$ can reflect the semantic relevance. The larger $H\left(S_v, P_v|S_h, P_v\right)$ is, the more irrelevant the visual and haptic sources are.

Initially, the coding efficiency of the general CMC is assessed, shown in Fig. 3.22. It illustrates the correlation between the bit error rate (BER) and the semantic $H\left(S_v, P_v|S_h, P_h\right)$ under a constant compression ratio. The variable L denotes the number of search paths within the decoding algorithm, indicating the complexity of the decoder. It is noticeable that its coding efficiency closely approximates the theoretical computational limit. The practical threshold at which the BER for S_v is nearly zero is $H\left(S_v, P_v|S_h, P_h\right) = 0.439bit$, contrasting with the theoretical limit of $0.5bit$. Subsequently, the rate-distortion curves for the theoretical general CMC model, the practical CMC codec, and the theoretical baseline model are compared, highlighting the superiority of the general CMC. When the two sources exhibit correlation ($H\left(S_v, P_v|S_h, P_h\right) = 0.5$), both the theoretical and practical rates of $R_{v,500}$ for a 500-length binary sequence in general CMC are significantly lower than the theoretical baseline model's value, demonstrated in Fig. 3.23. From these observations, two conclusions can be drawn. On one hand, since the practical values tend to converge with the theoretical values, the theorems proposed in Sect. 3.4.2 can be

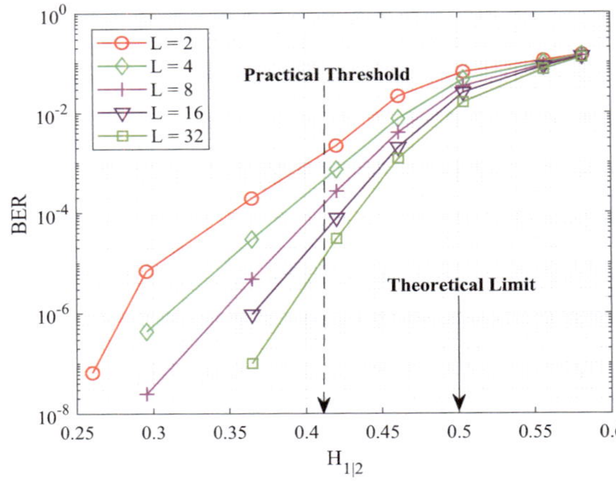

Fig. 3.22 Theoretical validation about coding efficiency

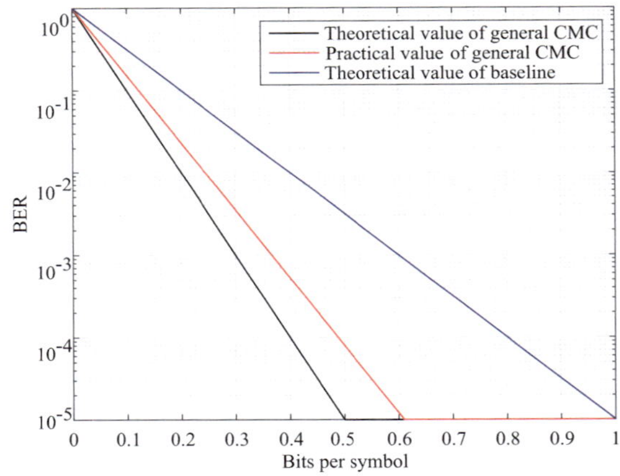

Fig. 3.23 Theoretical validation about rate distortion

verified. On the other hand, the general CMC achieves a theoretical benefit of 50% and a practical benefit of 37.5% over the baseline model if visual and haptic sources are semantically correlated.

Performance evaluation and analysis

The proposed general cross-modal codec in Sect. 3.4.3 is implemented on the developed teleoperation platform (shown in Fig. 3.12). Figure 3.24a–b show the original visual and haptic signals, and the corresponding predicted signals obtained from the cross-modal prediction. In order to evaluate the performance of the cross-modal prediction, Fréchet inception distance (FID) and RMSE between the original and the corresponding prediction signals are selected as criteria. Specifically, a priority-based scheduling is utilized for the transmission of diverse visual and haptic streams. To

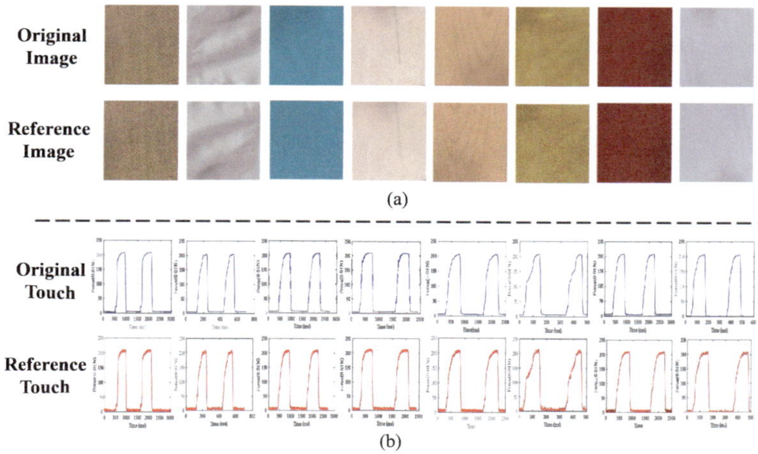

Fig. 3.24 Comparison between original data and prediction data generated from cross-modal prediction: **a** Images generated from haptic signals; **b** Haptic signals generated from images

gain an in-depth understanding, an analysis is employed for the comparative performance between CMVC and CMHC against the other competitive coding techniques.

Performance of CMVC: To assess the efficacy of the CMVC, we conduct tests on the rate and distortion of visual streams in comparison with widely used visual codecs (e.g., VVC, HEVC, and H.264). Within this experiment, rate performance is gauged through bits per pixel (bpp) and visual throughput, whereas distortion performance is quantified using peak signal-to-noise ratio (PSNR). Figure 3.25 presents a comparison of rate-distortion curves for CMVC and the competing VVC, HEVC, H.264 approaches, demonstrating that CMVC excels in coding efficiency, particularly under constrained rate conditions. In scenarios where rate resources are plentiful, the performance of the CMVC is constrained by the precision of the cross-modal predicted signal. The findings depicted in Fig. 3.26 clearly illustrate a substantial decrease in visual throughput facilitated by CMVC, while maintaining identical fidelity requirements.

Performance of CMHC: To confirm the advantages of CMHC, the end-to-end latency and PSNR of haptic streams are evaluated in comparison with the perceptual deadband (PD) method. Based on the data presented in Figs. 3.27 and 3.28, the latency of haptic streams is significantly lowered under identical transmission scheduling due to the substantial reduction in haptic data volume achieved by CMHC. Moreover, CMHC markedly enhances the reliability of haptic streams with PSNR$\geq$ 51 dB. The reason is that the CMHC effectively harnesses the semantic correlation across modalities to boost the fidelity of haptic signals.

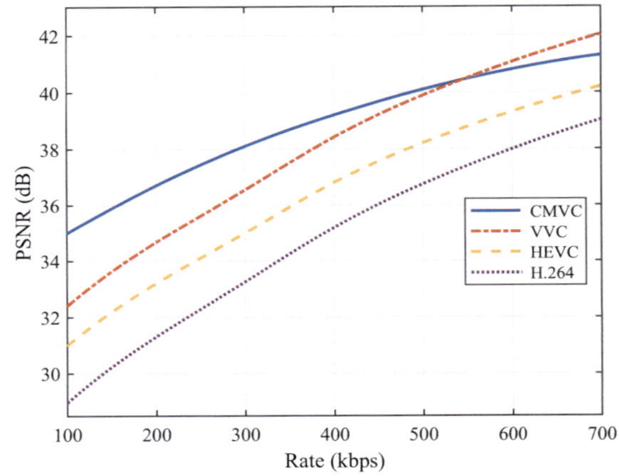

Fig. 3.25 Performance comparison about visual rate distortion

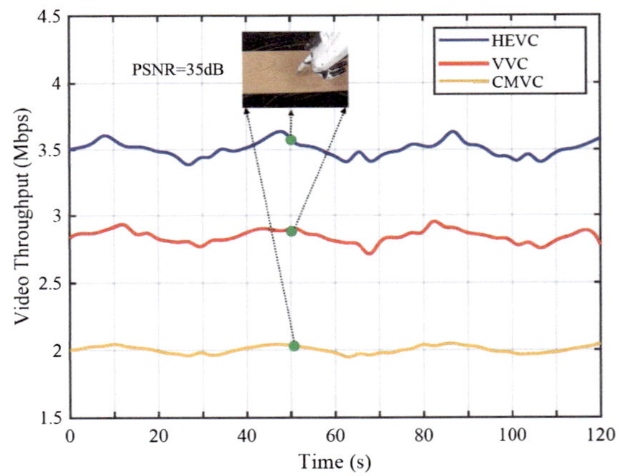

Fig. 3.26 Performance comparison about visual throughput

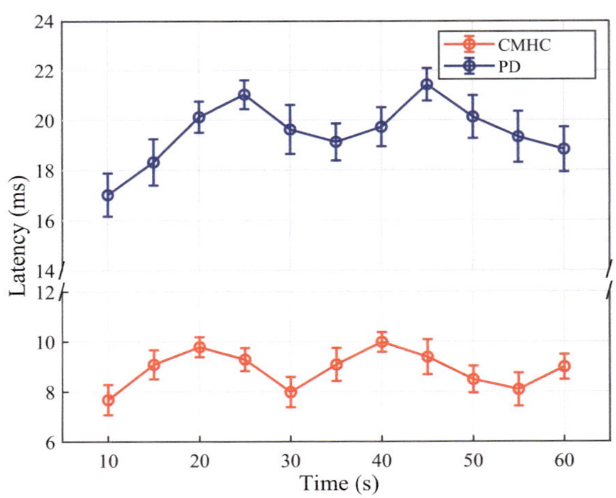

Fig. 3.27 Performance comparison about haptic latency

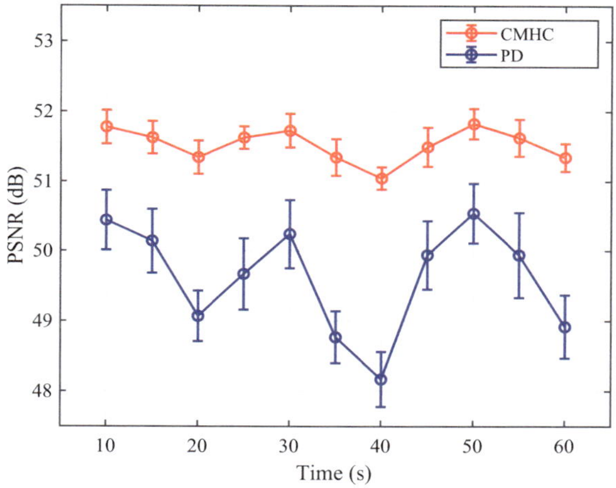

Fig. 3.28 Performance comparison about haptic reliability

3.5 Summary

In this chapter, we have described cross-modal coding techniques from two perspectives. On the one hand, we propose two approaches for content redundancy elimination. This includes a visual-haptic mutual redundancy elimination scheme and an audio-haptic joint redundancy elimination scheme. The aim is to try to compress the content of one modality's signals with the aid of the other modalities. The core idea is to leverage the correlations among modalities coming from the contents themselves or human's perception. From the experimental results, these approaches effectively reduce the content to be delivered and save communication resources. On the other hand, we also propose two approaches for coding to support both signal compression at the sender and signal reconstruction at the receiver. This includes a visual-aided haptic coding approach and a visual-haptic mutual coding approach. In these two approaches, we define the required variables and derive the important theorems using Shannon's information theory. The main difference is that we carefully consider the semantic correlations among modalities. For the former, the semantic entropy is defined, and the minimum number of bits required to compress haptic signals under the rate conditions of visual streams is explored. A practical cross-modal codec is also designed, which verifies the correctness of the associated theorems and approaches the theoretical limits. For the latter, an achievable rate region is defined, and the optimal source coding rate can be approximated by computing the infimum of this region. Under the guidance of this theory, a general cross-modal codec is constructed with the help of AI semantic analysis and polar code. Experimental results evaluate the performance of this general cross-modal codec. Importantly, the schemes introduced in this section can also be extended and applied to modalities other than visual and haptic signals by exploring the characteristics of those target modalities.

References

1. Wei X, Liao J, Zhou L, Sari H, Zhuang W (2024) Toward generic cross-modal transmission strategy. IEEE Trans Commun 72(10):6059–6072
2. Matlin MW, Foley HJ (1996) Sensation and perception. Allyn & Bacon, USA
3. Cizmeci B, Xu X, Chaudhari R, Alt N, Steinbach E (2017) A multiplexing scheme for multimodal teleoperation. ACM Trans Multim Comput 13(2):21
4. Yi X, Yang K, Zhao X, Wang Y, Yu H (2019) AHCM: adaptive Huffman code mapping for audio steganography based on psychoacoustic model. IEEE Trans Inf Foren Sec 14(8):2217–2231
5. Wei X, Yao Y, Wang H, Zhou L (2023) Perception-aware cross-modal signal reconstruction: from audio-haptic to visual. IEEE Trans Multimedi 25:5527–5538
6. Otsu N (1979) A threshold selection method from gray-level histograms. IEEE Trans Syst Man Cyb 9(1):62–66
7. Sullivan GJ, Ohm J-R, Han W-J, Wiegand T (2012) Overview of the high efficiency video coding (HEVC) standard. IEEE Trans Circ Syst Vid 22(12):1649–1668
8. Noll A, Gülecyüz B, Hofmann A, Steinbach E (2020) A rate-scalable perceptual wavelet-based vibrotactile codec. In: IEEE haptics symposium, Crystal City, USA, Mar 28–31, pp 854–859
9. Yuan Z, Kang B, Wei X, Zhou L (2022) Exploring the benefits of cross-modal coding. IEEE Trans Circ Syst Vid 32(12):8781–8794
10. Weaver W (1949) The mathematics of communication. Sci Am 181(1):11–15
11. Cover TM, Thomas JA (2006) Elements of information theory. Wiley, Hoboken, NJ, USA
12. Liveris AD, Xiong Z, Georghiades CN (2002) Compression of binary sources with side information at the decoder using LDPC codes. IEEE Commun Lett 6(10):440–442
13. Wei X, Shi Y, Zhou L (2022) Haptic signal reconstruction for cross-modal communications. IEEE Trans Multimedi 24:4514–4525
14. Yuan Z, Wu D, Zhou L (2024) Achieving the optimum rate for cross-modal source coding. IEEE Trans Multimedi 26:9722–9735
15. Cover TM, Thomas JA (2006) Elements of information theory. Wiley, USA
16. Bross B, Wang Y-K, Ye Y, Liu S, Chen J, Sullivan GJ (2021) Overview of the versatile video coding (VVC) standard and its applications. IEEE Trans Circ Syst Vid 31(10):3736–3764
17. Arikan E (2011) Systematic polar coding. IEEE Commun Lett 15(8):860–862
18. Wei X, Zhang M, Zhou L (2022) Cross-modal transmission strategy. IEEE Trans Circ Syst Vid 32(6):3991–4003

Chapter 4
Cross-Modal Transmission

Abstract In this chapter, the cross-modal transmission techniques are described in detail, which can be divided into two main categories. Schemes in the first category belong to the preemptive-resume scheduling pattern, where haptic streams have higher priority than audio-visual streams during transmission. On one hand, low-latency and high-reliability demands for haptic streaming should be guaranteed. Considering this, the intra-modal correlation-based haptic streaming scheme is proposed, which explores correlation within the haptic modality to employ efficient transmission. On the other hand, due to the occasional arrival of haptic streams, the throughput of audio-visual streams may well be reduced or even interrupted. The inter-modal correlation-based audio-visual streaming and content-aware stream scheduling schemes are respectively designed, handling the fluctuant audio-visual transmission quality issues with the aid of the haptic modality. Schemes in the second category belong to the non-preemptive-resume scheduling pattern. In other words, the transmission resources allocated to haptic and audio-visual modalities are under comprehensive and coordinated consideration. Among them, the modal-aware stream scheduling scheme flexibly settles the transmission priority of the modal stream instead of the data flow, achieving a tradeoff between various requirements. The adaptive prediction-aware scheme predicts and transmits the multi-modal signals in advance to reduce the whole latency during stream delivery. The dynamic transmission mode selection scheme tries to improve resource utilization efficacy by optimizing both the chosen transmission strategy and the allocated resource. Finally, the edge intelligence-empowered transmission decision scheme carefully considers factors such as communication, caching, computation, and control capacities, and realizes autonomous transmission decisions through AI technology.

4.1 Intra-Modal Correlation-Based Haptic Streaming

For haptic streaming, the delay and error in the signals using existing schemes make it extremely difficult to guarantee the continuity of the received haptic signals. In particular, these two factors are always mutually contradictory, bringing an extraordinary challenge under the scenarios with constrained wireless resources. Please

X. Wei et al., *Cross-Modal Communication Technology*, Wireless Networks,
https://doi.org/10.1007/978-3-032-00957-9_4

note that error probability is taken as the assessment of reliability, which refers to the likelihood of packet errors or packet loss. Furthermore, obtaining closed-form expressions for delay and reliability is challenging due to the bursty nature of haptic stream packets. Considering this, a joint transmission framework combining prediction and device-to-device (D2D) communication by leveraging intra-modal correlation of the haptic signals is proposed [1]. From a theoretical perspective, to entirely eradicate discontinuities, we advocate for a predictive mechanism that involves forecasting and preemptively dispatching forthcoming signals. To counterbalance the potential decline in reliability due to predictive errors, we efficiently establish D2D connections on the receiving end. From a technical end, we initially craft an algorithm that minimizes transmission resource usage, employing a binary search method to pinpoint the optimal prediction horizon, and then, we introduce a straightforward yet potent transmission mode selection algorithm.

4.1.1 System Model and Delay Analysis

The system model is shown in Fig. 4.1. The multi-modal streams, mainly composed of haptic streams, are transmitted from the senders to the receivers. There exist two transmission modes, where the multi-modal streams can be either transmitted from the senders to the receivers or assisted by the auxiliary robots (ARs) at the master domain. The former can be denoted as direct transmission, while the latter can be realized by D2D transmission. Suppose there is one sender, L ARs, and N receivers. The network domain encompasses two base stations (BSs) along with a core network, serving as the conduit for transmission between the senders and the receivers, as well as the transmission between the senders and the ARs. Particularly, it is noted that the values of L and N are determined by the requirement of the specific multi-modal services and network capacity. Meanwhile, ARs should be deployed around the receivers. The sets of the receivers and ARs are respectively denoted as $\mathcal{G} = \{g_1, \ldots, g_l, \ldots, g_L\}$ and $\mathcal{M} = \{m_1, \ldots, m_i, \ldots, m_N\}$. An AR is assumed to handle the multi-modal stream only for one receiver. To minimize the end-to-end delay, the senders anticipate the forthcoming haptic streams and preemptively dispatch them to the receivers. Concurrently, to leverage the benefits of D2D communications, the senders convey preliminary details of the haptic streams to these ARs, thereby enhancing their predictive capabilities.

Under this system model, the delay should be analyzed and derived. Here, time is discretized into slots, while the duration of each slot is T_s. The delay requirement is represented as the maximum latency bound or slots $D_{\max}$. The haptic signal to be transmitted from the sender to the receiver m_i in the kth slot is denoted as $X_{m_i}(k) = \left[X_{m_i}^1(k), \ldots, X_{m_i}^F(k)\right]^T$. F is the number of the related haptic features. The haptic signal that the receiver m_i gets in the kth slot is denoted as $Y_{m_i}(k)$. For the two transmission modes, $D_{\max}$ is the same. However, the calculation of delay in these two scenarios is completely different, which will be described in detail in the following.

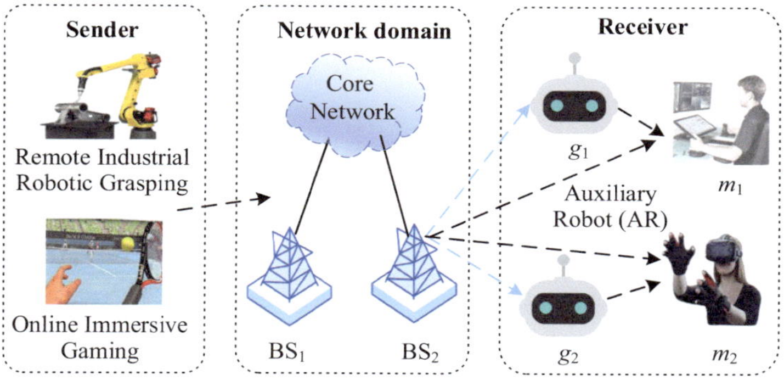

Fig. 4.1 System model

For *direct transmission*, multi-modal streams only experience the core network and the backhaul delay. Let $D^c_{m_i}$ denote the end-to-end delay from the sender to receiver m_i, which is also the human experienced delay. When no extra prediction mechanism is adopted, the relationship can be written as $Y_{m_i}(k + D^c_{m_i}) = X_{m_i}(k)$, shown in Fig. 4.2. Here, $D^c_{m_i}$ is the sum of the queuing delay $D^{q1}_{m_i}$ at the sender, the transmission delay $D^{t1}_{m_i}$ at the sender, the decoding delay $D^{d1}_{m_i}$ at BS1, the backhaul delay $D^r_{m_i}$ from BS1 to BS2, the transmission delay $D^{t2}_{m_i}$ at BS2, and the decoding delay $D^{d2}_{m_i}$ at receiver m_i. Importantly, let $T^p_{m_i}$ denote the prediction horizon, which means the reduction of delay by forecasting and preemptively dispatching the forthcoming haptic signal. As the transmission via the network domain is performed by optical fiber, the transmission delay and queuing delay are omitted. Therefore, the human experienced delay $D^{e1}_{m_i}$ should satisfy the condition in Eq. (4.1),

$$D^{e1}_{m_i} = D^c_{m_i} - T^p_{m_i} \leq D_{\max}. \tag{4.1}$$

For *D2D transmission*, the multi-modal stream of a receiver is delivered by an AR, shown in Fig. 4.3. The human experienced delay $D^{e1}_{m_i}$ should satisfy the condition in Eq. (4.2),

$$D^{q2}_{m_i} + D^{t3}_{m_i} + D^{d3}_{m_i} - T^p_{m_i} \leq D_{\max}, \tag{4.2}$$

where $D^{q2}_{m_i}$ is the queuing delay at the AR. $D^{t3}_{m_i}$ is the transmission delay at the AR to receiver m_i. $D^{d3}_{m_i}$ is the decoding delay at receiver m_i (Fig. 4.2).

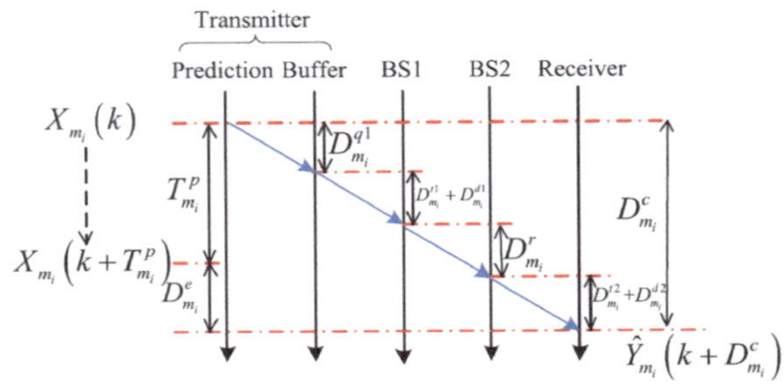

Fig. 4.2 Illustration of the prediction structure at the sender [1]

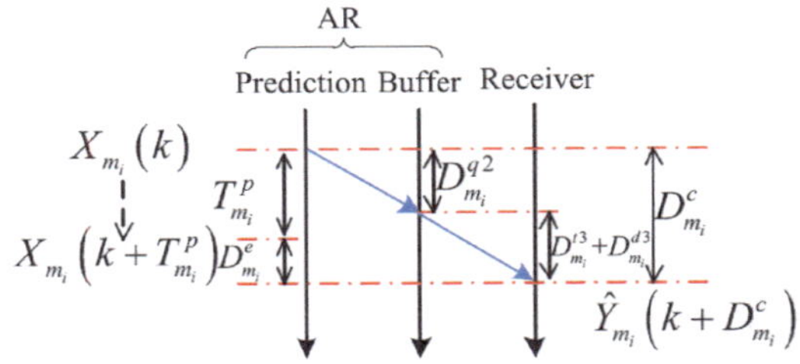

Fig. 4.3 Illustration of the prediction structure at an AR [1]

4.1.2 Reliability or Error Probability Analysis

Under this system model, the reliability should also be analyzed and derived. Here, the reliability demand is described by the maximum error probability $\varepsilon_{\max}$. For the two transmission modes, $D_{\max}$ is the same. However, the calculation of delay in these two scenarios is completely different, which will be described in detail in the following.

For *direct transmission*, the total error probability is contingent upon the prediction error probability $\varepsilon_{m_i}^{p1}$ of the sender, the transmission error probability $\varepsilon_{m_i}^{t1}$ from the sender to BS1, the transmission error probability $\varepsilon_{m_i}^{t2}$ from BS2 to the receiver, and the probability of queuing delay violation $\varepsilon_{m_i}^{q1}$. In the direct transmission scenario, the total reliability $\varepsilon_{m_i}^{d}$ can be derived by

$$\varepsilon_{m_i}^{d} = 1 - (1 - \varepsilon_{m_i}^{p1})(1 - \varepsilon_{m_i}^{q1})(1 - \varepsilon_{m_i}^{t1})(1 - \varepsilon_{m_i}^{t2}). \tag{4.3}$$

Due to the small value of all the probabilities ($\leq 10^{-4}$), $\varepsilon_{m_i}^{d}$ should satisfy the condition in Eq. (4.4),

$$\varepsilon_{m_i}^{d} \approx \varepsilon_{m_i}^{p1} + \varepsilon_{m_i}^{q1} + \varepsilon_{m_i}^{t1} + \varepsilon_{m_i}^{t2} \leq \varepsilon_{\max}. \tag{4.4}$$

For *D2D transmission*, the total error probability can be derived from the prediction error probability $\varepsilon_{m_i}^{p2}$ at the AR, the queuing delay violation probability $\varepsilon_{m_i}^{q2}$, and the transmission error probability $\varepsilon_{m_i}^{t3}$ from the AR to receiver. It should satisfy the condition in Eq. (4.5),

$$\varepsilon_{m_i}^{p2} + \varepsilon_{m_i}^{q2} + \varepsilon_{m_i}^{t3} \leq \varepsilon_{\max}. \tag{4.5}$$

In general, we have $\varepsilon_{m_i}^{p2} > \varepsilon_{m_i}^{p1}$ for the following reasons. Firstly, the predictive capability of an AR is inferior to that of the sender. Secondly, the sender must initially deliver information regarding the transmission process to the AR at regular intervals. This enables the AR to anticipate future haptic signals. It is inevitable that some errors will arise in the transmission link between the sender and the ARs.

Next, the closed-form expressions of the prediction error probability, queuing delay bound violation probability, and transmission error probability in Eqs. (4.4) and (4.5) are further derived. It is noted that they are respectively denoted as $\varepsilon_{m_i}^{p}$, $\varepsilon_{m_i}^{q}$, and $\varepsilon_{m_i}^{t}$, making no distinction between the above two transmission modes.

Prediction error probability analysis

At the receiver m_i, $X_{m_i}(k)$ changes by the following transition

$$X_{m_i}(k+1) = \Phi_{m_i} X_{m_i}(k) + W_{m_i}(k), \tag{4.6}$$

where $\Phi_{m_i} = [\phi_{m_i}^{f,f'}]_{F\times F}$, $f, f' = 1, 2, \ldots, F$ is the transition matrix associated with the specific transmission process. $W_{m_i}(k) = [w_{m_i}^{f}(k)]$, $f = 1, 2, \ldots, F$ is the transition noise, with each element $w_{m_i}^{f}(k)$ following Gaussian distribution about zero mean and variance $\sigma_{p,f}^2 = 0$.

Similarly, for predicting the future haptic signal, a linear model is considered here [2]. To be more specific, the haptic signal in the $(k+1)$th slot can be predicted as

$$\tilde{X}_{m_i}(k+1) = \Phi_{m_i} X_{m_i}(k) + V_{m_i}(k), \tag{4.7}$$

where $V_{m_i}(k) = [v_{m_i}^{f}(k)]$, $f = 1, 2, \ldots, F$ is the prediction noise. Its element follows a Gaussian distribution about zero mean and variances $\sigma_{p,f}^2$. For the direction transmission mode, the sender implements prediction and $\sigma_{p,f}^2 = 0$, while for the D2D transmission mode, an AR employs prediction, $\sigma_{p,f^2} > 0$. The different ARs have distinctive $\sigma_{p,f^2} > 0, \forall f = 1, \ldots, F$.

From Eqs. (4.6) and (4.7), the difference between the predicted haptic signal $\tilde{X}_{m_i}\left(k + T_{m_i}^{p}\right)$ and the real haptic signal $X_{m_i}\left(k + T_{m_i}^{p}\right)$ are

$$E_{m_i}\left(k + T_{m_i}^{p}\right) = X_{m_i}\left(k + T_{m_i}^{p}\right) - \tilde{X}_{m_i}\left(k + T_{m_i}^{p}\right). \tag{4.8}$$

The jth element of $E_{m_i}\left(k + T_{m_i}^{p}\right)$ is

$$\begin{aligned} e_{m_i}^{j}\left(k + T_{m_i}^{p}\right) =& w_{m_i}^{j}\left(k + T_{m_i}^{p} - 1\right) - v_{m_i}^{j}\left(k + T_{m_i}^{p} - 1\right) \\ &+ \sum_{i=1}^{T_{m_i}^{p}-1} \sum_{f=1}^{F} \phi_{m_i,j,f,T_{m_i}^{p}-i} \times \left[w_{m_i}^{f}(k+i-1) - v_{m_i}^{f}(k+i-1)\right], \end{aligned} \tag{4.9}$$

where $\phi_{m_i,j,f,T_{m_i}^p-i}$ is the element of $\left(\Phi_{m_i}\right)^{T_{m_i}^p-i}$ at the jth row and the fth column. It can be derived that $e_{m_i}^j\left(k+T_{m_i}^p\right)$ follows a Gaussian distribution with zero mean and variance $\rho_{m_i,j}^2\left(T_{m_i}^p\right)$,

$$\rho_{m_i,j}^2\left(T_{m_i}^p\right)=\sigma_{t,j}^2+\sigma_{p,j}^2+\sum_{i=1}^{T_{m_i}^p-1}\sum_{f=1}^{F}\left(\phi_{m_i,j,f,T_{m_i}^p-i}\right)^2\times\left(\sigma_{t,f}^2+\sigma_{p,f}^2\right). \quad (4.10)$$

Here, the just-noticeable difference (JND) of the haptic signal, represented as $\Delta=\left[\delta_1,\ldots,\delta_F\right]^T$, is utilized to constrain $e_{m_i}^j\left(k+T_{m_i}^p\right)$,

$$\begin{aligned}P\left\{\left|e_{m_i}^j\left(k+T_{m_i}^p\right)\right|\leqslant\delta_j\right\}&=1-P\left\{\left|e_{m_i}^j\left(k+T_{m_i}^p\right)\right|>\delta_j\right\}\\&=1-2\varphi\left(\frac{-\delta_j}{\rho_{m_i,j}\left(T_{m_i}^p\right)}\right),\end{aligned} \quad (4.11)$$

where $\varphi(\cdot)$ is the cumulative distribution function of the standard Gaussian distribution with zero mean and unit variance. Finally, for a haptic signal with F features, $\varepsilon_{m_i}^p$ can be obtained as

$$\varepsilon_{m_i}^p=1-\prod_{j=1}^{F}\left[1-2\varphi\left(\frac{-\delta_j}{\rho_{m_i,j}\left(T_{m_i}^p\right)}\right)\right], \quad (4.12)$$

which is related to $T_{m_i}^p$ under a specific JND.

Queuing delay bound violation probability analysis

The expression of $\varepsilon_{m_i}^q$ can be derived with the help of the concept of effective bandwidth [3],

$$\varepsilon_{m_i}^q=e^{D_{m_i}^q\phi\left(\lambda_{m_i},E_{m_i}^B\right)}, \quad (4.13)$$

where

$$\begin{aligned}\phi\left(\lambda_{m_i},E_{m_i}^B\right)&=E_{m_i}^B\mathbb{W}_{-1}\left(-\frac{\lambda_{m_i}}{E_{m_i}^B}e^{-\lambda_{m_i}/E_{m_i}^B}\right)+\lambda_{m_i},\\E_{m_i}^B&=\frac{\ln\left(1/\varepsilon_{m_i}^q\right)}{D_{m_i}^q\ln\left\{\left[\ln\left(1/\varepsilon_{m_i}^q\right)\right]/\left[\lambda_{m_i}D_{m_i}^q\right]+1\right\}}.\end{aligned} \quad (4.14)$$

$E_{m_i}^B$ is the effective bandwidth. $\mathbb{W}_{-1}(\cdot) < -1$ is the "–1" branch of the lambert function. λ_{m_i} (packets/slot) is the packet arrival rate of the haptic signals.

Transmission error probability analysis

For guaranteeing latency and reliability demands for haptic streaming, retransmissions without feedback can be considered here. Let $D_{m_i}^{\tau}$ denote the duration of each transmission, then $D_{m_i}^{\tau} = D_{m_i}^{t}/K_{m_i}$. $D_{m_i}^{t}$ is the transmission delay and K_{m_i} is the number of retransmissions. As haptic signals are transmitted with short packets, the error probability of each transmission is

$$\varepsilon_{m_i}^{\tau} = \int_0^{\infty} f_Q\left\{\sqrt{\frac{D_{m_i}^{\tau} B_{m_i}}{V_{m_i}}}\left[\ln\left(1 + \frac{a_{m_i} g_{m_i} p_{m_i}}{N_0 B_{m_i}}\right) - \frac{b_{m_i}\ln 2}{D_{m_i}^{\tau} B_{m_i}}\right] \cdot f_g(x)\right\} dx, \quad (4.15)$$

where $f_Q(\cdot)$ denotes the Q-function. B_{m_i} is the bandwidth. $V_{m_i} = 1 - \left[1 + \gamma_{m_i}\right]^{-2}$. $\gamma_{m_i} = a_{m_i} g_{m_i} p_{m_i}/N_0 B_{m_i}$ is the received SNR. a_{m_i} is the large-scale gain. g_{m_i} is the small-scale gain. p_{m_i} is the transmit power, and different transmission links correspond to different transmit powers. N_0 is the single-sided spectral density of additive Gaussian white noise. b_{m_i} is the bits in a block. $f_g(x)$ is the distribution of the instantaneous channel gain. Finally, $\varepsilon_{m_i}^{t}$ can be obtained as $\varepsilon_{m_i}^{t} = \left(\varepsilon_{m_i}^{\tau}\right)^{K_{m_i}}$.

4.1.3 *Problem Formulation and Optimization*

In cross-modal transmission, transmission power is an essential resource and critical factor affecting delay and reliability performance. Moreover, the prediction horizon plays a key role in meeting the delay requirements. In this context, we try to minimize wireless power consumption by optimizing the prediction horizon and transmission mode. To be more specific, the power consumption of the BS2 during direct transmission and D2D transmission modes should be carefully considered. In the case of D2D transmission, the relative positions of the ARs and receivers, as well as the predictive capabilities of the ARs, vary. Different combinations of ARs and receivers will result in varying power consumption levels. To address this, a binary variable x_{g_l,m_i} is defined to represent the transmission mode for receiver m_i. If $x_{g_l,m_i} = 1$, it indicates that the received haptic stream at m_i is delivered by AR g_l. Conversely, if $x_{g_l,m_i} = 0$, $\sum_{g_l \in \mathcal{G}} x_{g_l,m_i} = 0$ signifies that the received haptic stream is using direct transmission. Consequently, the problem can be formulated as follows:

$$
\begin{aligned}
\min_{T^p_{m_i}, x_{g_l,m_i}} \quad & \sum_{m_i \in \mathcal{M}} \sum_{g_l \in \mathcal{G}} x_{g_l,m_i} \tilde{p}_{m_i}, \\
\text{s.t. } & C1: p^{BS2}_{m_i} \le p^{\max}_{BS}, \quad \forall m_i \\
& C2: p^{AR}_{m_i} \le p^{\max}_{AR}, \quad \forall m_i \\
& C3: \sum_{g_l \in \mathcal{G}} x_{g_l,m_i} \le 1, \quad \sum_{m_i \in \mathcal{M}} x_{g_l,m_i} \le 1, \\
& \forall m_i, \forall g_l, \\
& (4.1)\ (4.2)\ (4.4)\ (4.5),
\end{aligned} \tag{4.16}
$$

where

$$
\tilde{p}_{m_i} = \begin{cases} p^{m_i}_{BS2}, & \sum_{g_l \in \mathcal{G}} x_{g_l,m_i} = 0 \\ p^{m_i}_{g_l}. & x_{g_l,m_i} = 1 \end{cases} \tag{4.17}
$$

$p^{m_i}_{BS2}$ and $p^{m_i}_{g_l}$ are respectively the transmission power of BS2 and AR g_l. Constraints $C1$ and $C2$ are the maximum transmission power constraints, where $p^{\max}_{BS}$ and $p^{\max}_{AR}$ are the transmission power thresholds of the BS and AR. Constraint $C3$ indicates that a receiver's stream can be handled by no more than one AR. Conversely, an AR can be responsible for delivering the stream to no more than one receiver. Provided that the transmission power is correctly configured, it is possible to meet the delay and reliability specifications for the two transmission modes. Owing to the benefits of the D2D transmission mode for communications over short distances, the energy consumption in D2D transmission is expected to be less than that in direct transmission, given an identical probability of transmission error.

The optimization for the problem about Eq. (4.16) is challenging because it concerns integer and $0 - 1$ variables to be handled, as well as the non-convex conditions. Based on this, it can be further divided into two subproblems to be optimized: prediction horizon optimization and transmission mode optimization.

Prediction horizon optimization

As the prediction horizon optimization is similar under the two transmission modes, the direct transmission mode is taken for describing the optimization process, that is

$$
\begin{aligned}
\min_{T^p_{m_i}} \quad & p^{m_i}_{BS2}, \\
\text{s.t. } & p^{m_i}_{BS2} \le p^{\max}_{BS}, \\
& (4.1)\ (4.4).
\end{aligned} \tag{4.18}
$$

In other word, the aim of this subproblem is to explore the minimal power consumption satisfying the delay and reliability demands, represented as $p^{m_i}_{BS2}(\varepsilon^{d,o}_{m_i}, D^{e1,o}_{m_i})$. Due to the difficulty of deriving the closed-form expression $p^{m_i}_{BS2}(\varepsilon^{d,o}_{m_i}, D^{e1,o}_{m_i})$, the minimization of $\varepsilon_{m_i}^{d,o}$ for a given $p_{BS2}^{m_i}$ should be firstly performed. Under this condition, this can be written to minimize $\varepsilon_{m_i}^{d,o}$ with respect to $T_{m_i}^{p}$

$$\begin{aligned} \min_{T_{m_i}^{p}} \ & \varepsilon_{m_i}^{p1} + \varepsilon_{m_i}^{q1} + \varepsilon_{m_i}^{t1} + \varepsilon_{m_i}^{t2} \\ \text{s.t.} \ & p_{BS2}^{m_i} \leq p_{BS}^{\max} \\ & D_{m_i}^{e1} \leq D_{\max} . \end{aligned} \tag{4.19}$$

Then, the minimal power consumption satisfying the reliability demand should be found, that is

$$\begin{aligned} \min_{p_{BS2}^{m_i}} \ & p_{BS2}^{m_i}, \\ \text{s.t.} \ & \hat{\varepsilon}_{m_i}^{d,\min} \left(p_{BS2}^{m_i}\right) \leq \varepsilon_{\max}. \end{aligned} \tag{4.20}$$

As the transmission error probability diminishes with $p_{BS2}^{m_i}$, power consumption is minimized when the constraint reaches equality. To simplify the complexity, the optimal solution can be derived using a binary search method. The specific solution for Eqs. (4.19) and (4.20) can be found in [1].

Transmission mode optimization

Drawing from prediction horizon optimization, a $(L+1) \times N$ matrix $\mathbf{P} = (p_{li})$, $l = 1, \ldots, L+1, i = 1, \ldots, N$ is obtained. To be more specific, for $l = 1, \ldots, L$, $p_{li} = p_{g_l}^{m_i}$, and for $l = L+1$, $p_{li} = p_{BS2}^{m_i}$. Subsequently, the transmission mode optimization subproblem is formulated as

$$\begin{aligned} P2: \min_{x_{g_l,m_i}} \ & \sum_{m_i \in \mathcal{M}} \sum_{g_l \in \mathcal{G}} x_{g_l,m_i} p_{li}, \\ \text{s.t.} \ & \sum_{g_l \in \mathcal{G}} x_{g_l,m_i} \leq 1, \quad \sum_{m_i \in \mathcal{M}} x_{g_l,m_i} \leq 1, \\ & \forall m_i, \quad \forall g_l. \end{aligned} \tag{4.21}$$

To tackle the above subproblem, a straightforward yet potent scheme based on the Hungarian algorithm can be employed, boasting polynomial-time computational complexity [4]. However, several issues and the corresponding adjustments should be carefully considered.

Firstly, due to the assignment properties of the Hungarian algorithm, it might conclude that a sender can handle at most one receiver's stream, which is inadequate in multi-modal services. To resolve this, the elements indicating power consumption associated with senders delivering haptic streams to receivers can be removed. Secondly, when an AR cannot transmit a receiver's haptic stream, a high constant value can be assigned to represent its power consumption. If no ARs are capable of delivering a particular receiver's stream, the traditional Hungarian method might still assign an AR. To address this, the power consumption data for that receiver can be eliminated, ensuring it is delivered by direct transmission from the sender. Thirdly, if the number of ARs is fewer than the number of receivers, the traditional Hungarian algorithm may fail to match some receivers with an AR. In such situations, the matrix to square dimensions can be adjusted, with the unassigned receivers' streams being delivered by direct transmission from the sender. Fourthly, the solution obtained may not be the most efficient. If a receiver's stream is delivered by an AR, the power consumption may be higher compared to direct transmission from the sender. To optimize this, the two transmission modes should be compared and the one with the lower power consumption is selected as the final solution.

4.2 Inter-Modal Correlation-Based Audio-Visual Streaming

In the existing preemptive-resume scheduling strategy, a haptic packet with high priority interrupts the transmission of the low-priority audio-visual packet. The transmission of the low-priority audio-visual packet is resumed after transmission of the high-priority haptic packet [5]. Therefore, the throughput of audio-visual streams is seriously impacted by the prioritized haptic streams, which further reduces the quality of the received audio-visual signal and human experience. Particularly, due to the bursty characteristic of haptic streams, existing 5G network slicing techniques become unavailable [6, 7]. Based on this, for enhancing the throughput of audio-visual streaming, an inter-modal correlation-based audio-visual streaming is designed, suitable for the preemptive-resume scheduling strategy [8]. Initially, the arrival times of bursty haptic streams are predicted by inter-modal correlation between them and audio-visual modalities. And then, rather than pausing the transmission of audio-visual streams, the remaining network resources are allocated to them using edge computing and network slicing techniques. Figure 4.4 illustrates the difference between the traditional scheme and the proposed scheme in this section.

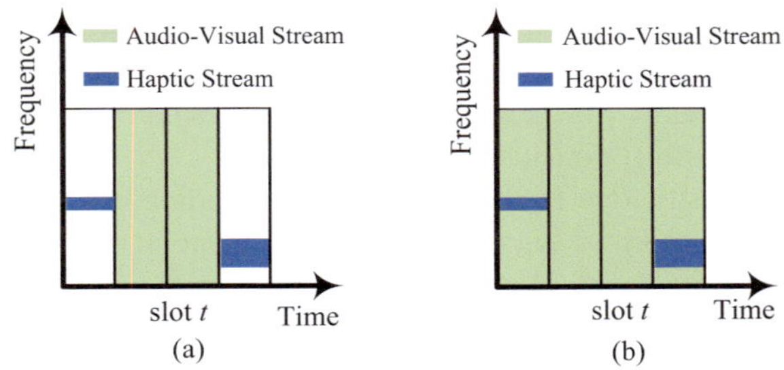

Fig. 4.4 Illustrations of **a** the traditional scheduling scheme and **b** the proposed scheduling scheme [8]

4.2.1 *Audio-Visual-Aided Haptic Arrival Time Estimation*

To precisely estimate the arrival time of the bursty haptic streams, the core concept is to leverage the inherent spatiotemporal correlations between streams of different modalities. The main process is shown in Fig. 4.5. Firstly, a rough interval about the arrival time of the next haptic signal is analyzed based on a vast array of historical haptic signals stored in the database. Subsequently, the exact arrival time within this rough interval can be swiftly pinpointed by examining the spatial correlations between frames in real-time audio-visual streams.

Rough interval of haptic arrival time analysis

When compared to the continuous nature of audio-visual streams, the bursty haptic streams are relatively sparse. Consequently, it should incorporate haptic signals exhibiting similar temporal patterns from other pertinent terminals in the communication network for a distributed analysis. However, from an application standpoint, this often raises concerns about the communication expenses associated with the data sharing mechanism [9]. Particularly, since a large volume of haptic signals is directly pooled from various terminals, the technical challenge at hand is to achieve distributed estimation of this interval within the confines of communication limitations.

Specifically, this problem can be solved by computing the average across various time series, where the temporal indices with non-zero values can fully encompass the precise arrival time of the haptic signals. Hence, considering the time series of haptic streams, which includes T discrete time stamps, $h_z \in [0, 1]^T$ for each record z. The objective is to compute the mean across a total of $\mathcal{Z}$ records from accessible terminals, that is $\bar{h} = \frac{1}{\mathcal{Z}} \sum_z h_z$, while adhering to the b-bit communication constraint. To accomplish this objective, it can further involve steps about data preprocessing, quantization, and sampling.

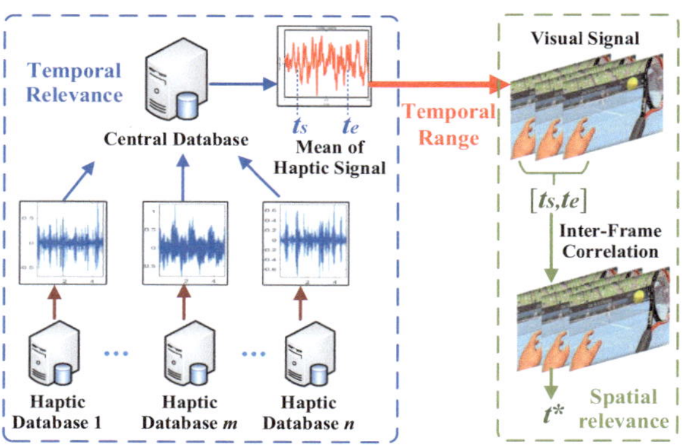

Fig. 4.5 Spatiotemporal relevance analysis for haptic arrival time

Data preprocessing: Given the uneven distribution of information in h_z, for example, $h_z = \{0, 0.2, \ldots, 0.1\}$, the alternative parameters can be utilized to replace h_z for the sake of communication efficiency. To be more specific, to bolster resistance to noise, it is crucial that h_z is evenly distributed across these transmitted parameters. To achieve this, h_z can be preprocessed using the orthogonal basis $e_j \in \mathbb{R}^T$, that is $h_z = \sum_{j=1}^{N} \rho_j e_j$, where this is originated from the Kashin's representation. ρ_j instead of h_z can reflect parameters fulfilling $\max_j |\rho_j| \leq \frac{O}{\sqrt{N}} \|h_z\|_2$. As a result, h_z can be constrained with $\{\rho_j\}_{j=1}^{N} \in [-\frac{r}{\sqrt{T}}, \frac{r}{\sqrt{T}}]^{cT}$, where r and c are constants.

Quantization: Considering the communication constraint, ρ_j should be confined within the b-bit limit through quantization. To be more specific, upon receiving $\rho_j \in [-\frac{r}{\sqrt{T}}, \frac{r}{\sqrt{T}}]$, this parameter is quantized into 1-bit q_j, that is $q_j \in \{-\frac{r}{\sqrt{T}}, \frac{r}{\sqrt{T}}\}$. The total bits required for h_z can be capped at $N = \Theta(T)$. Regarding q_j, the following quantization rules are established to ensure an unbiased estimation of ρ_j, $\mathbb{E}\left[q_j\right] = \rho_j$,

$$q_j = \begin{cases} -\frac{r}{\sqrt{T}}, & \text{with probability } \frac{r/\sqrt{T}-\rho_j}{2r/\sqrt{T}} \\ \frac{r}{\sqrt{T}}, & \text{with probability } \frac{r/\sqrt{T}+\rho_j}{2r/\sqrt{T}} \end{cases}. \tag{4.22}$$

Sampling: To further reduce the communication cost from N to b bits, b bits are randomly sampled from $q = \left\{q_1, \ldots, q_j, \ldots, q_N\right\}$,

$$U = \left(q_{u_1}, \ldots, q_{u_b}\right) \in \left\{-\frac{r}{\sqrt{T}}, \frac{r}{\sqrt{T}}\right\}^b, \tag{4.23}$$

where $u_1, \ldots, u_b$ are index randomly selected from N. In summary, considering U, the subsequent estimation can be derived, $\mathbb{E}\left(N \cdot q_{u_m} \cdot \mathbb{I}\{j = u_m\}\right) = \mathbb{E}\left[q_j\right] = \rho_j$. Define $\check{\rho}_j = \frac{N}{b} \sum_{m=1}^{b} q_{u_m} \cdot \mathbb{I}\{j = u_m\}$, and thus an unbiased estimate of the mean $\check{h}$ can be obtained by using the "Kashin's representation". With the result $\check{h}$, the time interval $[t_s, t_e]$ for the haptic signal's arrival is also determined experimentally by identifying the non-zero indices.

Exact haptic arrival time determination

After estimating the rough interval of haptic arrival time, the next aim is to determine the exact arrival time. Here, audio-visual information can provide key information on whether the haptic-related event such as touching the target object happens or not. Consider this, the touched target object is firstly parsed from visual signals, which can eliminate redundant information. This can be realized by the attention mechanism. Then, inter-frame correlations can be computed to find the key frame of the touched target object. To be more specific, let V_t denote the visual signal in frame t. The inter-frame correlation $\tau(t-1, t)$ between frame $t-1$ and t is computed, that is $\tau(t-1, t) = V_{t-1} \cdot V_t$. Meanwhile, the intraframe correlation in frame $t-1$

is computed, that is $\tau(t-1) = V_{t-1} \cdot V_{t-1}$. The loss function is defined as

$$\Upsilon_{loss}(V_{t-1}, V_t) = \frac{|\tau(t-1,t) - \tau(t-1,t-1)|}{\tau(t-1,t-1)} - \vartheta, \tag{4.24}$$

where ϑ is the threshold depending on the concrete multi-modal services. Finally, within the rough interval, the exact haptic arrival time $t^\star$ can be determined. It corresponds the key haptic frame where $\Upsilon_{\text{loss}}(t^\star - 1, t^\star) > 0$.

4.2.2 Resource Allocation for Audio-Visual Streaming

In light of the aforementioned arrival times of haptic streams, a stream scheduling scheme can be performed. The aim is to allocate appropriate resources for haptic streams with high priority at time $t^\star$, while also providing resources for audio-visual streams as much as possible. Specifically, audio-visual streams $A\&V$ are deployed across slots, which can be adaptively split into $\mathcal{M}$ minislots to accommodate streams. Haptic streams H are occupied at minislot and each minislot is represented by $\hbar(m)$, with m being an element of $\mathcal{M}$. $\hbar$ can be calculated as the sum of all $\hbar(m)$ across the minislots, and the expectation of $\hbar$ is denoted as $\mathbb{E}(\hbar) \triangleq \rho$.

Due to fluctuations in the wireless channels, the slot at each state $\omega \in \mathcal{W}$ is treated as an independent and identically distributed process [10]. Essentially, the combined scheduling strategy Π is designed to determine the suitable allocation of network resources to the multi-modal streams under channel state ω for each minislot m. Consequently, within the scope of minislot m for channel state ω, the scheduling for H with high priority is represented as

$$\Pi_H = \left[\Omega_m^\omega\right]_{\omega \in \mathcal{W}, m \in \mathcal{M}}, \tag{4.25}$$

where Ω represents the fraction of resources assigned to H. Accordingly, $\rho \triangleq \mathbb{E}(\hbar) = \mathbb{E}(\Omega^\omega)$.

Based on this, the remaining resources are allocated to $A\&V$ to avoid interrupting their transmission, thereby increasing their throughput, as shown in Fig. 4.4. This is the main difference from the traditional scheme such as [5]. Specifically, the stream scheduling for $A\&V$ at each minislot m and a given channel state ω is

$$\Pi_{A\&V} = \left[\Psi_m^\omega\right]_{\omega \in \mathcal{W}, m \in \mathcal{M}} = 1 - \Omega_m^\omega, \tag{4.26}$$

where Ψ represents the allocated remaining resources.

From the above analysis, the joint scheduling strategy $\Pi = \{\Pi_H, \Pi_{A\&V}\}$ can be optimized by maximizing data rates R^ω at the channel state ω,

$$R^\omega = \Xi\left(\Pi_H, \Pi_{A\&V}, r_m^\omega\right). \tag{4.27}$$

$\Xi(\cdot)$ represents the rate allocation function, which usually adopts a linear model [11]. r_m^{ω} is the preset peak rate of each minislot [7]. Furthermore, Eq. (4.27) can be broken down into two simpler components:

$$R_H^{\omega} = r^{\omega}\mathbb{E}\left(\Omega^{\omega}\right) = r^{\omega}\mathbb{E}(\hbar) = \sum_m r_m^{\omega}\rho_m,$$
$$R_{A\&V}^{\omega} = \sum_m r_m^{\omega}\left(1 - \rho_m\right). \tag{4.28}$$

R_H^{ω} and $R_{A\&V}^{\omega}$ are respectively the rate resources allocated to haptic and audio-visual streams. From Table 2.1, the demanded haptic streaming capacity is small. Therefore, at the arrival time of haptic streams, approximately 200 Kbps is sufficient for haptic modality, while the remaining resource can be given for guaranteeing audio-visual streaming.

In order to evaluate the effectiveness of this inter-modal correlation (IMC)-based audio-visual streaming scheme, audio-visual throughput is adopted. It represents data rates of audio-visual streams being transmitted from the slave domain to the master domain. The result is presented in Fig. 4.6. Compared to the haptic communications paradigm (HCP) [12], the IMC scheme has significantly improved the throughput of audio-visual streams. It is evident that, despite the presence of high-priority haptic streams, their influence on audio-visual streams is considerably less pronounced than in HCP. The reason is that the allocation of network resources for audio-visual streams is carefully coordinated without disrupting their transmission.

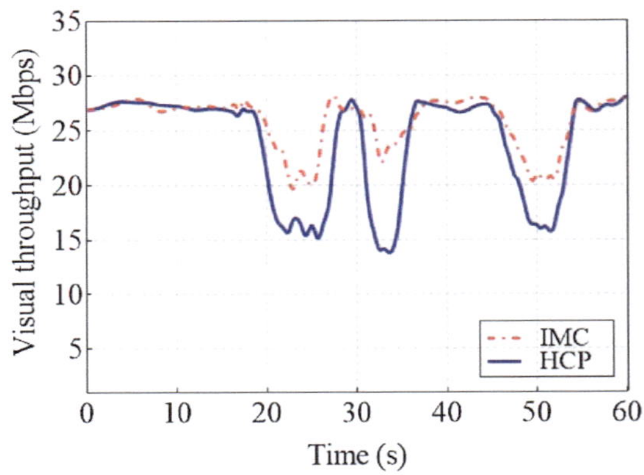

Fig. 4.6 Performance comparison in terms of throughput

4.3 Content-Aware Stream Scheduling

Besides using the scheme described in Sect. 4.2 to alleviate the throughput reduction about audio-visual streams in the preemptive-resume scheduling strategy, another idea is to adopt AI-driven cross-modal signal recovery techniques, which will be introduced in Sect. 4.5. That is, the damaged or delayed audio and visual signals that have semantic correlation with haptic signals can be generated from the haptic modality at the receiver. However, the effectiveness of the recovery often hinges on the content correlation between the haptic signal and the damaged portions of the audio-visual signals. For example, in tele-diagnosis, the remote physician must evaluate the patient's status using real-time haptic and visual inputs. The visual signal serves primarily to monitor both the medical robot and the patient. The visual stream that captures the medical robot's actions portrays the same subject from another perspective as the haptic feedback signal. In other words, there exists a correlation between haptic and this visual stream. Conversely, the visual stream that focuses on the patient does not share this correlation with the haptic feedback signal. If the patient-focused visual stream is interrupted by the haptic signal and becomes compromised, it cannot be effectively recovered at the receiver. Regrettably, this situation is probable because, in systems that prioritize haptic signals, the haptic data stream could randomly interrupt any ongoing visual stream. Therefore, a scheme for content-aware cross-modal stream transmission that utilizes content correlations to link haptic preemptive scheduling with visual signal recovery is proposed [13]. To begin with, a general cross-modal stream transmission issue is established as the maximization of visual utility subject to constraints imposed by haptic requirements. Subsequently, an online algorithm for content-aware resource allocation is developed to address the streaming challenge by dynamically preempting visual streams that have a high correlation with haptic streams.

4.3.1 System Model and Problem Formulation

System model

We examine a representative multi-modal service such as tele-diagnosis. Here, a haptic stream denoted as h is to be transmitted in conjunction with several visual streams $\mathcal{V} = \{v_1, v_2, \ldots, v_{|\mathcal{V}|}\}$. Time is segmented into slots of fixed length. Within each slot, there is a collection of minslots $\mathcal{N} = \{1, 2, \ldots, |\mathcal{N}|\}$. The maximum resources that can be allocated to the multi-modal services in a slot are $f_{\max}$. In the context of preemptive-resume scheduling, visual stream scheduling is conducted across different slots, while the haptic stream is scheduled within minislots upon visual streams. The demand for haptic packets, denoted by A_n, in each minislot n adheres to a Poisson distribution with a mean rate of λ. For each channel state $s \in \mathcal{S}$, a multi-modal streaming scheme, π, can be characterized by the following elements.

(1) A resource allocation method for visual stream is defined as $\xi = \{\xi^s_{v,n} | \xi^s_{v,n} = \xi^s_v / |\mathcal{N}|, \text{ and } \sum_{v \in \mathcal{V}} \xi^s_v \leq f_{\max}\}$. ξ^s_v signifies the resources designated for visual stream v during a slot in channel state s. Furthermore, the resources designated for v in each minislot n is $\xi^s_{v,n}$.

(2) A haptic adaptive queuing system permits each incoming haptic packet to wait in a queue for minislots ϖ that can be adjusted based on actual latency needs.

(3) A haptic preemptive scheduling is denoted as $\delta = \{\delta^s_{v,n} | v \in \mathcal{V}, n \in \mathcal{N}, s \in \mathcal{S}\}$. $\delta^s_{v,n}$ indicates the haptic stream h_n interrupting the communication resources $\xi^s_{v,n}$ designated for v in minislot n and channel state s.

(4) A visual signal recovery procedure has a rate enhancement factor $g(\gamma_{h,v})$. It describes the effect of reconstruction on v, contingent on the content correlation between haptic and visual modalities $\gamma_{h,v}$. Here, $\gamma_{h,v}$ is determined through deep learning techniques. To elaborate, a convolutional neural network (CNN) and a recurrent neural network (RNN) are initially used to extract features from the visual and haptic signals, respectively. These features are then combined and integrated before being input into a fully connected neural network. The aforementioned model is trained with the Pearson correlation coefficient serving as the performance metric. Ultimately, with a trained end-to-end deep learning-based model, the sender is able to ascertain the content correlation among visual and haptic signals to be transmitted.

Haptic requirement constraint

Within the queuing system, if a haptic data packet has arrived but is not scheduled for transmission in the upcoming ϖ minislot, it is discarded. The carrier of these haptic packets is essentially an $M/M/S$ queuing model designed for users who have a limited waiting tolerance of ϖ. The waiting tolerance can be represented as a stochastic variable characterized by its cumulative distribution function $G(\varpi)$. The average outage probability of the haptic stream is specified by

$$P_{out} = \sum_{s \in \mathcal{S}} p_S(s)(1 - (1 - \alpha)(1 - p^e_s)), \tag{4.29}$$

where $p_S(s)$ is the distribution of channel state s. p^e_s is the transmission error probability. α is the haptic packet abandonment rate. Here, each haptic packet with L-bits is allocated bandwidth b in each minislot. Based on [14], p^e_s is calculated as

$$p^e_s = Q\left(\frac{bA_n \times C(SNR_s) - LA_n}{\sqrt{bA_n \times V(SNR_s)}}\right), \tag{4.30}$$

where $Q(\cdot)$ is the Gaussian Q-function, SNR_s is the signal-to-noise ratio, $C(SNR_s) = \log_2(1 + SNR_s)$ and $V(SNR_s) = 1 - \frac{1}{(1+SNR_s)^2}$. On one hand, the expected steady-state queue length $\mathbb{E}[\mathcal{Q}]$ should be as short as possible, which is closely related to the low-latency demand. On the other hand, P_{out} should be less than a threshold ϑ for satisfying the high-reliability requirement. That is,

$$P1 : \min_{\lambda,\rho} \mathbb{E}[\Omega],$$
$$s.t. \;\; P_{out} \leq \vartheta, \; b\lambda\rho \leq f_{\max}, \tag{4.31}$$

where ρ denotes the haptic served rate, and $b\lambda\rho \leq f_{\max}$ constrains the allocatable resources.

Visual utility maximization

In this transmission scheme π, the rate of visual stream v at the sender, $r_v^{\pi,s}$, is determined by ξ_v^s and δ_v^s. In order to describe the fact that the visual stream is interrupted by the haptic stream, the rate allocation function $f(\cdot)$ and rate loss function $h(\cdot)$ [11]. The latter $h(\cdot)$ can model the visual stream rate loss influenced by the haptic stream. Therefore, $r_v^{\pi,s}$ is

$$r_v^{\pi,s} = f(\xi_v^s, \delta_v^s) = \frac{\xi_v^s}{f_{\max}} \breve{r}_v^s \left(1 - h\left(\frac{\delta_v^s}{\xi_v^s}\right)\right), \tag{4.32}$$

where $\breve{r}_v^s$ is the preset peak rate and $r_v^{\pi,s} \leq f(\xi_v^s, 0)$. When there is no haptic stream, the equality is satisfied. After considering signal recovery at the receiver, this rate can be rewritten as $r_v^{\pi,s} = f(\xi_v^s, \delta_v^s)g(\gamma_{h,v})$. Then, its expectation, defined as $\bar{r}_v^{\pi,s}$, is

$$\begin{aligned} \bar{r}_v^{\pi,s} &= \mathbb{E}\left[f(\xi_v^s, \delta_v^s)g(\gamma_{h,v})\right] \\ &= \xi_v^s \breve{r}_v^s g(\gamma_{h,v}) \left(1 - \mathbb{E}\left[h\left(\frac{\delta_v^s}{\xi_v^s}\right)\right]\right). \end{aligned} \tag{4.33}$$

Based on this, the throughput capacity of the visual stream v, $\bar{r}_v^{\pi}$, is obtained by the average value over a long duration,

$$\bar{r}_v^{\pi} = \sum_{s \in \mathcal{S}} \bar{r}_v^{\pi,s} p_S(s), \; \forall v \in \mathcal{V}. \tag{4.34}$$

If visual requirements exceed $\bar{r}_v^{\pi}$, the visual streams may be lost. The achievable rate region $\mathcal{R}$ for visual streams under a scheme π is defined as

$$\mathcal{R} = \{\bar{r}_v \mid \forall v \in \mathcal{V}, \exists\, \pi \text{ such that } \bar{r}_v \leq \bar{r}_v^{\pi}\}. \tag{4.35}$$

$\bar{r}_v = \sum_{s \in \mathcal{S}} \bar{r}_v^s p_S(s)$, whose unit is $Mbps$.

Moreover, a utility function $U_v(\bar{r}_v)$ should be defined to represent the satisfaction level of π for v within the constraints of haptic requirements. For example, $U_v = \log(\bar{r}_v)$ [11]. Finally, the objective is to identify an optimal scheme π^*, that maximizes $U_v(\bar{r}_v)$ over all the visual streams under constraint $P1$ and the following constraint,

$$P2: \max_{\pi} \sum_{v \in \mathcal{V}} U_v(\bar{r}_v),$$
$$s.t.\ P1\ and\ \bar{r}_v \in \mathcal{R}. \tag{4.36}$$

4.3.2 Content-Aware Cross-Modal Resource Allocation

In this section, we resolve the above problem using a content-aware cross-modal resource allocation (CCRA). To be specific, we begin by finding the optimal solution to the haptic requirement constraints $P1$. According to the fluid approximations for the $M/M/S$ queueing system, we have

$$\begin{aligned} &\mathbb{E}[\mathcal{Q}] \approx \lambda \int_0^{\bar{\varpi}} \frac{1}{(1+\varpi)^2} d\varpi, \\ &\alpha \approx \lambda(1-\rho)^+, \\ &\rho = \max\left\{ y^{-1}(0), \frac{f_{\max}}{b\lambda} \right\}, \end{aligned} \tag{4.37}$$

where $\bar{\varpi}$ is the mean patience time. $y(\rho) = P_{out} - \vartheta$, and $y^{-1}(\cdot)$ is the inverse function of $y(\rho)$. For the specific calculation of ρ, please refer to [15].

Here, the resources allocated to haptic streams in a minislot should be done proportionally based on pre-specified weights $\Omega = \{\omega_v^s(\gamma_{h,v}) | v \in \mathcal{V}, s \in \mathcal{S}\}$. Based on the core idea of this scheme and the existing cross-modal signal recovery techniques, the haptic stream can only occupy the resources of visual streams having high correlations in the content and good channel status. Therefore, the haptic puncturing weight $\omega_v^s(\gamma_{h,v})$ is determined by $\gamma_{h,v}$ and s, where $\sum_{v \in \mathcal{V}} \omega_v^s = 1$. Considering this, the haptic load on v under s and minislot n can be calculated as

$$\delta_{v,n}^s = \frac{\omega_v^s(\gamma_{h,v})}{\sum\limits_{v' \in \mathcal{V}} \omega_{v'}^s(\gamma_{h,v'})} b\mu_n = b\mu_n \omega_v^s(\gamma_{h,v}), \tag{4.38}$$

where $\mu_n = A_n \rho$. On one hand, the amount of haptic load puncturing on a visual stream should not exceed the resources allocated on that slot, $\delta_{v,n}^s \le \xi_{v,n}^s$. The condition in the following can be derived:

$$b\mu_n \le \min_{v \in \mathcal{V}} \frac{\xi_{v,n}^s}{\omega_{v,n}^s(\gamma_{h,v})}. \tag{4.39}$$

On the other hand, $\sum_{v \in \mathcal{V}} \delta_{v,n}^s \le f_{\max}/|\mathcal{N}|$, where $f_{\max}/|\mathcal{N}|$ is the maximum resources allocated to visual streams per minslot. Therefore,

$$b\mu_n = \sum_{v\in\mathcal{V}} \omega_v^s(\gamma_{h,v}) b\mu_n \leq \frac{f_{\max}}{|\mathcal{N}|}. \tag{4.40}$$

To guarantee Eq. (4.39), $\xi_{v,n}^s$ and $\omega_v^s(\gamma_{h,v})$ can fulfill the following relationship,

$$\frac{\xi_{v,n}^s}{\omega_v^s(\gamma_{h,v})} \geq \frac{f_{\max}}{|\mathcal{N}|},\ \forall v \in \mathcal{V}. \tag{4.41}$$

Based on this, the set of the CCRA can be described as

$$\Pi^{ca} = \left\{(\xi, \omega(\gamma)) \mid \text{and } \xi \geq \frac{f_{\max}}{|\mathcal{N}|}\omega(\gamma)\right\}. \tag{4.42}$$

Moreover, considering Eqs. (4.35) and (4.38), $\bar{r}_v^{\pi,s} = \mathbb{E}[f(\xi_v^s, \omega_v^s(\gamma_{h,v})\mu) g(\gamma_{h,v})]$. Based on Eq. (4.42), the achievable region constrained to CCRA can be represented as

$$\mathcal{R}^c = \left\{\bar{r}_v \mid \forall v \in \mathcal{V}, \exists\ (\xi, \omega) \in \Pi^{ca},\ \bar{r}_v \leq \bar{r}_v^{\pi}\right\}. \tag{4.43}$$

Next, the assumption is made: for all $s \in \mathcal{S}$ and $v \in \mathcal{V}$, the function $F_v^s(\cdot, \cdot)$ is defined as

$$F_v^s(\xi_v^s, \omega_v^s(\gamma_{h,v})) = \mathbb{E}\left[f\left(\xi_v^s, \omega_v^s(\gamma_{h,v})\mu\right) g(\gamma_{h,v})\right]. \tag{4.44}$$

It can be proved that $\mathcal{R}^c$ constrained to CCRA is convex [11, 13]. Here, $\omega_v^s(\gamma_{h,v})$ is abbreviated as ω_v^s. Under this convex characteristic, Eq. (4.36) about $P2$ can be rewritten under CCRA,

$$\begin{aligned} &\max_{(\xi_v^s, \omega_v^s)\in\Pi^{ca}} \mathcal{U}(\xi_v^s, \omega_v^s), \\ &\quad s.t.\quad \bar{\mathbf{r}}(\xi_v^s, \omega_v^s) \in \mathcal{R}^c, \end{aligned} \tag{4.45}$$

where $\mathcal{U}(\xi_v^s, \omega_v^s) = \sum_{v\in\mathcal{V}} U_v\left(\sum_{s\in\mathcal{S}} p_S(s) F_v^s(\xi_v^s, \omega_v^s)\right)$. Furthermore, in order to solve $P2$, Eq. (4.45) should be decomposed into two subproblems for the pair of coupling variables ξ_v^s and ω_v^s. To be more specific, when ω_v^s is fixed, the first subproblem is

$$\max_{\xi_v^s} \mathcal{U}(\xi_v^s, \omega_v^s) = \max_{\xi_v^s} \sum_{v\in\mathcal{V}} U_v\left(\sum_{s\in\mathcal{S}} p_S(s) F_v^s(\xi_v^s, \omega_v^s)\right). \tag{4.46}$$

The optimal solution can be obtained, denoted as ξ_v^{s*} and $\mathcal{U}^*$. Meanwhile, when ξ_v^{s*} is fixed, the second subproblem is

$$\max_{\omega_v^s} \mathcal{U}^*(\xi_v^{s*}, \omega_v^s) = \max_{\omega_v^s} \sum_{v\in\mathcal{V}} U_v\left(\sum_{s\in\mathcal{S}} p_S(s) F_v^s(\xi_v^{s*}, \omega_v^s)\right). \tag{4.47}$$

The above two subproblems can be solved by a gradient/subgradient algorithm. In the practical multi-modal services, only the channel state at the last round is available; an online gradient descent algorithm is adopted for utility maximization. Specifically, the utility maximization can be considered as a two-level iterative game between the scheduling decision maker and the dynamic environment. At round t,

$$\mathcal{U}_t(\xi_v^{s(t)}, \omega_v^{s(t)}) = \sum_{v\in\mathcal{V}} U_v\left((1-\kappa_t)\hat{r}_v^{t-1} + \kappa_t F_v^s(\xi_v^{s(t)}, \omega_v^{s(t)})\right). \tag{4.48}$$

The details are given in Algorithm 4.1, where $\mathcal{P}_{\Pi^{ca}}\{*\}$ represents the projection of $*$ onto Π^{ca}.

Algorithm 4.1 Procedure of utility maximization [13]

Require: CCMT Π^{ca}, iterations T, step sizes $\{\eta_t\}$
1: Set $t = 1, \kappa_t, (\xi_v^{s(t)}, \omega_v^{s(t)}) \in \Pi^{ca}$, and rate estimates $\hat{r}_v^t$, $v \in \mathcal{V}$
2: **for** $t = 1$ to T **do**
3: Compute $\mathcal{U}_t(\xi_v^{s(t)}, \omega_v^{s(t)})$
4: Let $\xi_v^{s(t+1)} = Prj_{\Pi^{ca}}\left\{\xi_v^{s(t)} - \eta_t \frac{\partial \mathcal{U}_t(\xi_v^{s(t)}, \omega_v^{s(t)})}{\partial \xi_v^{s(t)}}\right\}$
5: Compute $\mathcal{U}_t^*(\xi_v^{s(t+1)}, \omega_v^{s(t)})$
6: Let $\omega_v^{s(t+1)} = Prj_{\Pi^{ca}}\left\{\omega_v^{s(t)} - \eta_t \frac{\partial \mathcal{U}_t^*(\xi_v^{s(t+1)}, \omega_v^{s(t)})}{\partial \omega_v^{s(t)}}\right\}$
7: Update $\hat{r}_v^{t+1} = (1-\kappa_{t+1})\hat{r}_v^t + \kappa_{t+1} F_v^s(\xi_v^{s(t+1)}, \omega_v^{s(t+1)}(\gamma_{h,v}))$
8: **end for**
Output: ξ_v^{s*}, ω_v^{s*}, and $\hat{r}_v^*$

Finally, we apply the designed content-aware cross-modal transmission (CCMT) scheme to a constructed teleoperation platform shown in Fig. 3.12. The duration of each minislot is 0.125 ms and a slot has 8 minislots. The maximum resource allocated to multi-modal streams per slot is 120 RBs. The HEVC is applied for visual streams, while the perceptual deadband approach is utilized for haptic coding, similar to that in Sect. 3.3 [16]. The approach proposed in Sect. 5.2 is selected as the visual signal recovery scheme. The other system parameters are the same as those in Sect. 4.4 [17]. The performance of the CCMT is evaluated in terms of latency, reliability, and throughput. To draw a comprehensive observation, CCMT is benchmarked against two alternative transmission schemes: MMT [5] and CMT [17].

From Figs. 4.7, 4.8, and 4.9, the CCMT enhances the throughput of visual streams under high haptic service rates compared to the other two competing schemes, while maintaining minimal latency and optimal reliability for haptic streams. Specifically, as depicted in Fig. 4.7, the CCMT excels in reducing latency for visual streams and nearly matches CMT in terms of haptic stream latency. This is due to the adaptable

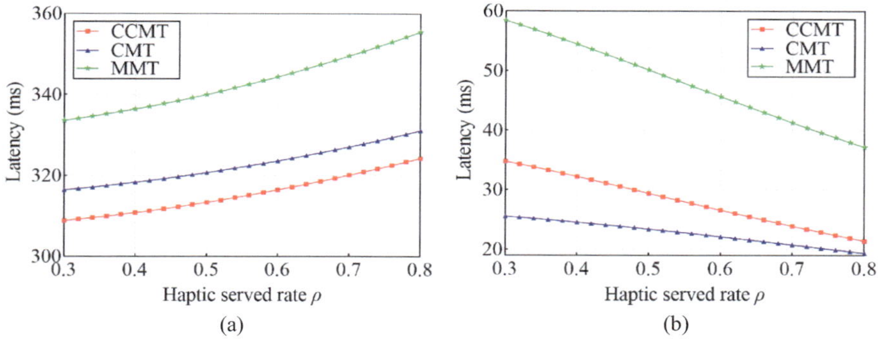

Fig. 4.7 Latency performance: **a** Visual stream; **b** Haptic stream

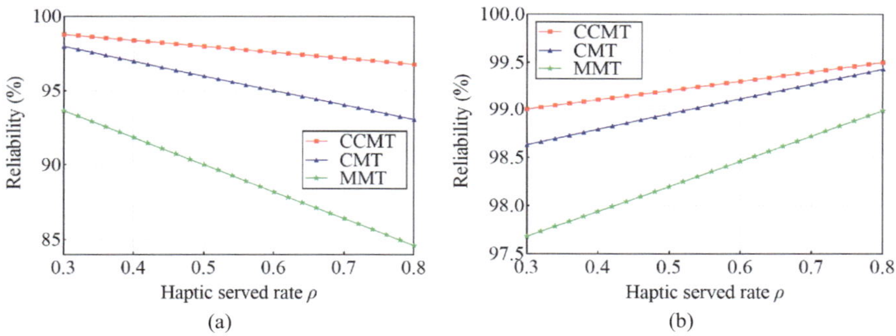

Fig. 4.8 Reliability performance: **a** Visual stream; **b** Haptic stream

queuing mechanism, which minimizes the latency impact of haptic streams on visual streams, albeit with a slight increase in latency. As the haptic service rate increases, the impact of latency on visual streams from haptic streams also rises progressively. Furthermore, Fig. 4.8 shows the reliability of visual and haptic streaming across the CCMT, CMT, and MMT. This indicates that an efficient queuing mechanism can mitigate the reliability decline when the haptic stream is not scheduled promptly at low haptic service rates. Lastly, Fig. 4.9 illustrates the visual throughput and haptic throughput for the CCMT, CMT, and MMT. The CCMT notably approaches the highest visual throughput and the least reduction in visual throughput as the haptic service rate ascends. Moreover, by capitalizing on correlations among modalities, the haptic throughput is decreased through a streamlined coding mechanism in both CCMT and CMT.

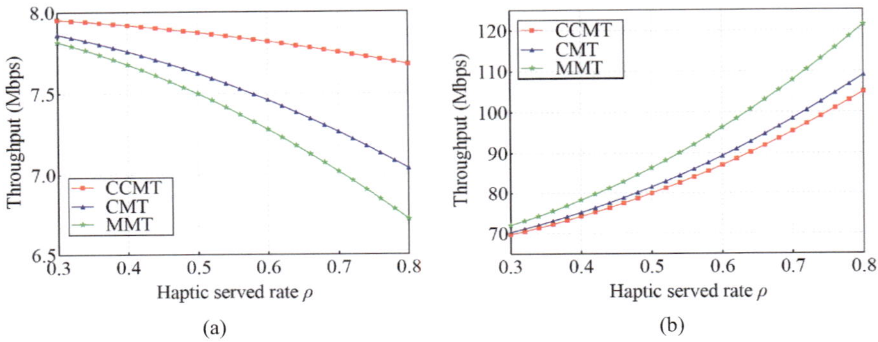

Fig. 4.9 Throughput performance: **a** Visual stream; **b** Haptic stream

4.4 Modal-Aware Stream Scheduling

In multi-modal services, audio, visual, and haptic streams have heterogeneous transmission requirements, such as low latency, high reliability, high throughput, and low complexity. Several requirements sometimes lead to contradictory actions. For example, in order to guarantee highly reliable transmission, the relatively computationally complicated coding and resource allocation methods should be adopted. In several preemptive-resume scheduling strategies, such as in Sects. 4.1, 4.2 and 4.3, setting priority for haptic stream is a feasible way. However, it will inevitably affect the transmission quality of audio-visual streams due to the dynamic puncturing from haptic modality and the dynamic network environment, further leading to human experience deterioration. Different from traditional data-driven or task-driven schedulers, this section presents a modality stream-driven scheduling paradigm [17]. It comprehensively makes use of transmission content, service requests, and human experience. Specifically, as described by Sect. 2.1, multi-modal services are initially categorized into three patterns: haptic-dominant, audio-visual-dominant, and equally important services. To accommodate these different kinds of multi-modal services, a suite of spatio-temporal-semantic aware streaming schemes is designed to handle their diverse requests. Furthermore, by refining the scheduling switch mechanism with a focus on semantic continuity, surprisingly, human experience can be significantly enhanced.

4.4.1 Problem Formulation and Design Principle

Here, $\mathcal{X}$ is defined as the set of multi-modal services. In these services, audio and visual streams are operated across the slot, while haptic streams are employed across the minislot, similar to those in Sects. 4.2 and 4.3. A slot usually comprises several minislots $\mathcal{N}$, which can be flexibly changed according to specific demands. The related demand in a minislot is denoted as $Q(n)$, $n \in \mathcal{N}$, and that in a slot is $Q \sim \sum_{n\in\mathcal{N}} Q(n)$. The mean and the cumulative function of Q are respectively $\mathbb{E}[Q] = \rho$ and C_Q. Moreover, in a slot, the wireless channel state is denoted as $s(s \in \mathcal{S})$, where $\mathcal{S}$ is the channel state set. The related channel capacity is $C_x^s(x \in \mathcal{X})$, which can be occupied by the specific multi-modal services. The network resource allocation for audio, visual, and haptic streams in minislot n, channel state s, and service x can be represented by a matrix $\omega \in \Omega$,

$$\Omega \triangleq \left\{ \omega \in \mathbb{R}_+ \middle| \max \sum_{x\in\mathcal{X}} \omega_{x,n}^s, \forall n \in \mathcal{N}, s \in \mathcal{S} \right\}. \tag{4.49}$$

$\omega_{x,n}^s$ and $\omega_x^s = \sum_{n\in\mathcal{N}} \omega_{x,n}^s$ respectively denote the allocated resources in the minislot and the related slot.

In the design of stream scheduling, heterogeneous transmission requirements on delay D, jitter J, data loss rate L, and data rate R should be considered. For the multi-modal services, scheduling can be further divided into resource allocation in a slot Ξ_d for audio-visual streams and that in a minislot Ξ_r for haptic streams. From this definition, the following principles should be obeyed.

Firstly, $\Xi_r = [\Xi_{r,s}^{x,n}]_{x\in\mathcal{X},n\in\mathcal{N},s\in\mathcal{S}}$, which means the fraction frequency-time slot resources allocated to service x at minislot n and channel state s. Its priority is higher than that of Ξ_d, especially in haptic-dominant services. The aim is to get low latency and high reliability, that is $\min_{D,J,L} \Xi_r(D, J, R, L)$. Secondly, $\Xi_d = [\Xi_{d,s}^{x,n}]_{x\in\mathcal{X},n\in\mathcal{N},s\in\mathcal{S}} = 1 - \Xi_r$. The aim is to maximize the data rate of audio-visual streams at s, R^s, that is $R^s = f^s(\Xi_r, \Xi_d)$. In other words, these data rates should be restricted by both Ξ_r and Ξ_d. Thirdly, Ξ_d and Ξ_r should be flexibly switched. When designing the switch mechanism, the factors in Ξ_r and Ξ_d, as well as the allocated resources described by Eq. (4.49), should be carefully considered.

When facing different service requests, signal contents, and transmission environments, transmission priority about haptic and audio-visual modalities should be adaptively updated. Moreover, the correlation among different modality signals is also considered here, which is the key difference between cross-modal transmission and traditional communications. The notation P is adopted to depict this correlation, containing spatiotemporal-semantic relevances.

4.4.2 Stream Scheduling Among Different Requests

After giving the definition of the key notations and the design principle, the stream scheduling for haptic-dominant services, audio-visual-dominant services, and equally important services are respectively described.

Haptic-dominant multi-modal service

In haptic-dominant services, the aims are minimizing the volume of the haptic stream and maximizing the network efficiency. It is essential to implement a resource multiplexing mechanism utilizing spatial diversity and multi-path transmission. To be more specific, firstly, the rotation angle ϕ_x is calculated as $\phi_x = \sum_s R^s \omega_x^s / \rho$. Then, from the original channel diversity $\phi_x \log \phi_x$, the additional channel diversity gain G_x can be calculated through bit interleaving and modulation angle rotation, that is $G_x = \phi_x \log \phi_x P / C_Q$. Subsequently, to accurately differentiate between random packet loss and congestion-induced packet loss, the rate control method presented in [18] is implemented. The appropriate number of paths N_p guaranteeing streaming quality under conditions of (D, J, R, L) is determined accordingly, $N_p = \lfloor \rho G_x / \omega_x^s f^s (D, J, R, L) \rfloor$. Finally, with the state parameters of the path and the delay constraints being recognized, Ξ_r for each potential path that can enhance the end-to-end delay and reliability volatility is

$$\Xi_r = \max_s \left\| R^s \omega_x^s \right\| \frac{\lfloor \rho G_x / \omega_x^s \rfloor}{f^s (D, J, R, L)}. \tag{4.50}$$

Audio-visual-dominant multi-modal service

In audio-visual-dominant services, the aims are to guarantee the transmission quality of large-scale audio-visual streams and relieve the influence of haptic interruption to the greatest extent. It is essential to implement virtual resource management of the network slicing, which contains three main phases for the virtualized network function (VNF).

In the *orchestrate phase*, a multi-dimensional and high-order VNF instance clustering problem is first established. It is represented as an evolutionary game $G(\Xi_r, \phi_x)$, which describes the compatibility among VNF instances. Then, this clustering is transformed into an evolutionary stability calculation,

$$G(\Xi_r, \phi_x) \sim \max_s \frac{\left\| R^s \omega_x^s \right\| R_c}{f^s (D, J, R, L) C_Q}. \tag{4.51}$$

Subsequently, the cluster numbers are computed by

$$N_c = \max_s \frac{\left\| R^s \omega_x^s \right\|}{f^s (D, J, R, L)} / \min_s \frac{\left\| R^s \omega_x^s \right\|}{f^s (D, J, R, L)}. \tag{4.52}$$

Finally, the related VNF can be represented as $G(\Xi_r, \phi_x) \sim G(N_c, \phi_x, f^s)$.

In the *deployment phase*, to guarantee low-latency and low-computational complexity, the VNF deployment should be replaced by the service function chain deployment. Firstly, the latter can be defined as a partial observation Markov decision process for adapting the dynamic network environment,

$$\max_{s} \frac{\left\lfloor \rho \left\| \phi_x \log \phi_x \right\| / R^s \omega_x^s \right\rfloor}{f^s\left(D, J, R, L\right)}. \tag{4.53}$$

Then, an approximate method is utilized to describe the dynamic service function chain deployment,

$$G\left(N_c, \phi_x, f^s\right) \sim G\left(\max_s \frac{\left\| R^s \omega_x^s \right\|}{f^s\left(D, J, R, L\right)}\right). \tag{4.54}$$

In the *arrangement phase*, to employ VNF scheduling, the VNF's execution sequence is firstly re-arranged by using the Markov decision process,

$$\max_{s} G\left(N_c, \left\| R^s \omega_x^s \right\|, f^s\right) \sim \max_{s} \frac{\left\lfloor \rho \left\| \phi_x \log \phi_x \right\| / R^s \omega_x^s \right\rfloor}{f^s\left(D, J, R, L\right)}. \tag{4.55}$$

Then, the virtual machine selection and VNF arrangement can be jointly optimized by

$$\max_{s} \frac{G\left(N_c, \left\| R^s \omega_x^s \right\|, f^s\right)}{f^s \rho \left\| \phi_x \log \phi_x \right\|}. \tag{4.56}$$

The greedy algorithm in [19] can be adopted to solve Eq. (4.55).

Equally important multi-modal service

In equally important services, mobile edge computing is considered as a promising technique. Specifically, dynamic stream priority adjustment is utilized to guide resource allocation, with the priority of the modal-aware stream M_x updated by

$$M_x\left(t+1\right) = \begin{cases} M_x\left(t\right) + \frac{\left\lfloor P \left\| \phi_x \log \phi_x \right\| / \rho C_Q \right\rfloor}{\max_s f^s\left(D, J, R, L\right)}, & \left\| \phi_x \right\| \le \rho, \\ M_x\left(t\right) - \frac{\left\lfloor P \left\| \phi_x \log \phi_x \right\| / \rho C_Q \right\rfloor}{\max_s f^s\left(D, J, R, L\right)}, & \left\| \phi_x \right\| > \rho. \end{cases} \tag{4.57}$$

Then, the selection and collaboration of multiple virtual machines under low latency and complexity constraints are

$$G\left(\Xi_r, \phi_x\right) \sim \lim_{t \to \infty} P_x\left(t\right) \max_{s} \left\| \frac{R^s \omega_x^s}{f^s\left(D, J, R, L\right)} \right\|. \tag{4.58}$$

This way can achieve the tradeoff between the high throughput and low fluctuation during streaming.

4.4.3 Scheduling Switch Mechanism

In a practice scenario, it is difficult to accurately determine the service pattern, especially for an online decision demand. Therefore, the scheduling switch mechanism is required for achieving performance smoothness. It can decide which scheduling scheme mentioned above should be implemented. Firstly, the possible reasons resulting in performance fluctuation need to be analyzed. Let the quality of service of the correct and incorrect modal-aware scheduling respectively denote as $QoS\left(\Xi_r, \Xi_d\right)$ and $\widehat{QoS}\left(\Xi_r, \Xi_d\right)$, the smooth of the performance S_{QoS} can be defined by the gap between these two items,

$$S_{QoS} = \frac{\max \left\| QoS\left(\Xi_r, \Xi_d\right) - \widehat{QoS}\left(\Xi_r, \Xi_d\right) \right\|}{QoS\left(\Xi_r, \Xi_d\right)}. \tag{4.59}$$

Assume different kinds of multi-modal services occur randomly, and the multi-modal streams are scheduled by schemes in Sect. 4.4.2, S_{QoS} fulfills

$$S_{QoS} \geq \frac{G\left(\phi_x P / C_{QoS}, \max_s \left\| R^s\left(t\right) \omega_x^s \right\|\right)}{\max_s f^s\left(D, J, R, L\right)}, \tag{4.60}$$

where C_{QoS} is the compensation function.

Based on the above definitions and analysis, the scheduling switch mechanism can be designed, containing three main steps. The first step is *timely judgement*. In this step, the concept of modal factor $f\left(t\right)$ is introduced, which changes the associated content analysis process $G\left(\phi_x\right)$, that is

$$G\left(\phi_x\right) \sim f\left(t\right) \max_s \left\| \frac{R^s\left(t\right) \omega_x^s}{f^s\left(D, J, R, L\right)} \right\|, \tag{4.61}$$

where $f\left(t\right)$ is updated by

$$f\left(t+1\right) = f\left(t\right) \frac{\left\lfloor P \left\|\phi_x \log \phi_x\right\| / \rho C_{QoS} \right\rfloor}{\max_s f^s\left(D, J, R, L\right)}. \tag{4.62}$$

In this way, the complexity of the algorithm can be reduced and accuracy can be guaranteed. The second step is *adaptive switch*. In this step, the adaptive switch method $A\left(\Xi_r, \Xi_d\right)$ is devised to minimize the fluctuation of the multi-modal streams,

$$A\left(\Xi_r, \Xi_d\right) = \max_t f\left(t\right) G\left(\phi_x, \max_s \left\| R^s\left(t\right) \omega_x^s \right\|\right). \tag{4.63}$$

Here, $G\left(\phi_x, \max_s \left\| R^s\left(t\right) \omega_x^s \right\|\right)$ acts as the metric to decide which scheduling is deployed. When

$$\|A(\Xi_r, \Xi_d)\| \geq \frac{\lfloor P \|\phi_x \log \phi_x\| / \rho C_{QoS} \rfloor}{\max_s f^s(D, J, R, L)}, \tag{4.64}$$

then $\Xi_r \rightarrow \Xi_d$, and $\Xi_d \rightarrow \Xi_r$ once

$$\|A(\Xi_r, \Xi_d)\| \leq \max_s \left\| \frac{R^s \omega_x^s}{f^s(D, J, R, L)} \right\|. \tag{4.65}$$

The third step is *reasonable compensation*, which can give a reference for the subsequent scheduling by readjusting the priority of the multi-modal streams. Based on Eq. (4.59), C_{QoS} can be configured as

$$C_{QoS} = \frac{S_{QoS} \|A(\Xi_r, \Xi_d)\|}{G(\phi_x)} \tag{4.66}$$

for reducing $\left\| QoS(\Xi_r, \Xi_d) - \widehat{QoS}(\Xi_r, \Xi_d) \right\|$ to the great extent.

Finally, the performance of the proposed modal-aware stream scheduling scheme is evaluated based on latency, reliability, throughput, and complexity. There are also two competing scheduling schemes to be considered, that are the content-aware scheduling (CAS) [20] and the task-driven scheduling (TDS) [7]. Specifically, the CAS selects distinct transmission methods based on the content being transmitted, whereas the TDS utilizes a rotation mechanism based on the completion cost of tasks. Three kinds of modality streams are sent randomly and can be received simultaneously. Figure 4.10 exhibits the performance of the proposed scheme with regard to latency, throughput, and reliability. By comparing the two competing schemes, the following conclusions can be drawn. On the one hand, there has been a significant reduction in the latency for haptic streams. Conversely, the latency for visual streams has not seen substantial changes. On the other hand, the variability in visual streams has been notably decreased. Moreover, the reliability of haptic streams has increased significantly. Notably, despite the random occurrence of haptic streams, their performance is on par with that of visual streams in terms of smoothness.

4.5 Prediction-Aware Stream Scheduling

In cross-modal transmission, low end-to-end delay demand is important and holds significant implications. Previous works on delay reduction mainly concentrate on priority-based or modal-aware scheduling, efficient codec design, etc. Besides these ways, adopting signal prediction techniques to compensate end-to-end delay for obtaining low-latency transmission is another effective route [2, 21]. The core concept involves predicting upcoming signals, including audio, visual, or haptic signals, and proactively transmitting them to the receiver. Consequently, the end-to-end delay is significantly mitigated by diminishing the perceived latency at the receiver's end.

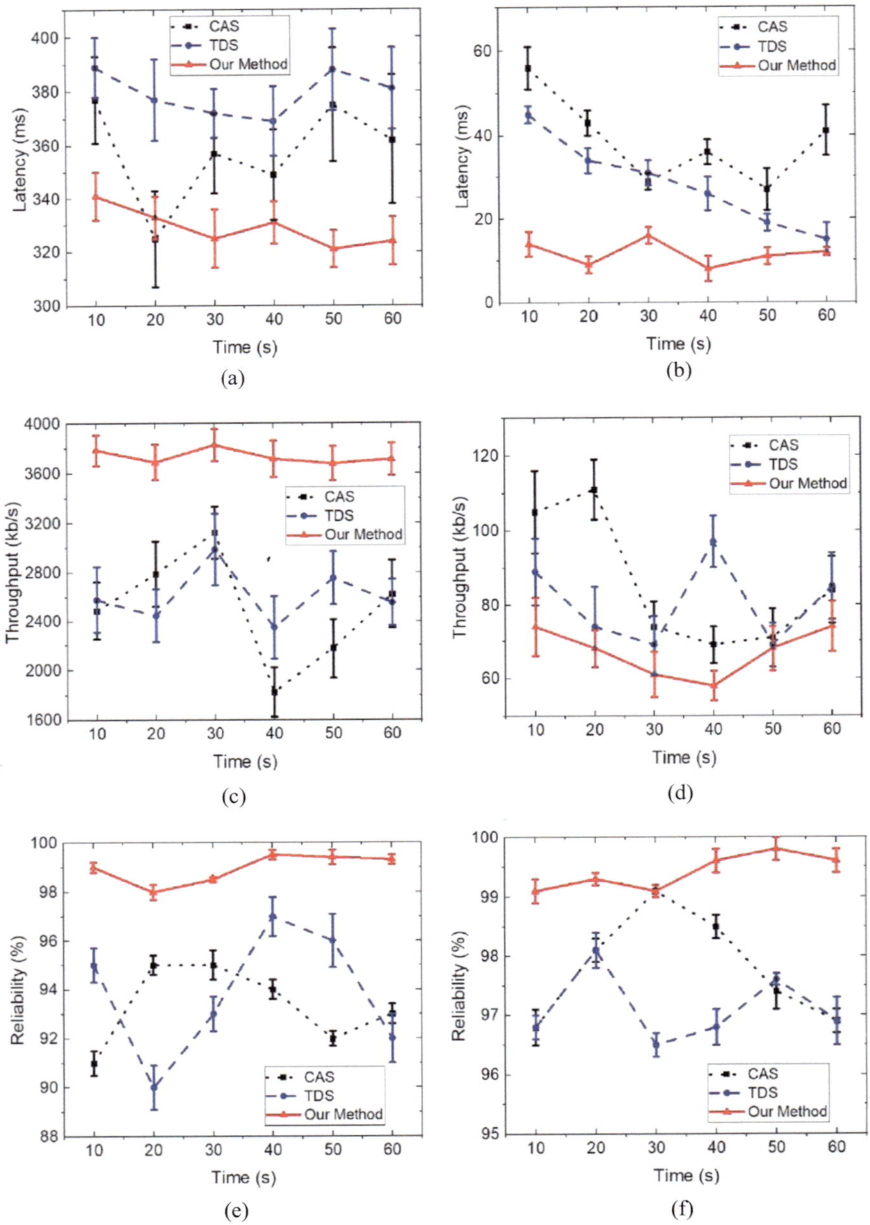

Fig. 4.10 Scheduling performance comparison: **a** Latency of visual stream; **b** Latency of haptic stream; **c** Throughput of visual stream; **d** Throughput of haptic stream; **e** Reliability of visual stream; **f** Reliability of haptic stream [17]

However, existing prediction solutions still face challenges. On one hand, the diversity of multi-modal services arises from the fundamental differences in the characteristics of multi-modal signals. It is necessary to construct an adaptive and flexible prediction framework for these services. On the other hand, correlation among modalities is used for realizing prediction, but the poor quality of feature extraction and optimization hinders the prediction accuracy, further impacting the transmission efficiency and user experience. To tackle these issues, this section provides a prediction-aware stream scheduling paradigm [22]. Specifically, the adaptive framework has been constructed for predicting the upcoming multi-modal signals and delivering them to the receiver in advance. The main characteristic is that it incorporates prediction feedback and supports prediction horizon adjustment under changing conditions. It also employs upcoming signal prediction in a cross-modal way and by the information gain-assisted graph attention (IGGA) method, which can improve the prediction accuracy and robustness.

4.5.1 Adaptive Prediction-Aware Streaming Framework

The framework for adaptive prediction-aware streaming is illustrated in Fig. 4.11. It can be divided into four components: adaptive prediction, communication network, error correction, and prediction feedback. The audio, visual, haptic signals at time t are respectively represented as A_t, V_t, and H_t. Firstly, the adaptive prediction component realizes the prediction of upcoming signals. Here, the historical window size is denoted as w and the prediction horizon is denoted as k. The signal at time $t+k$ is predicted by the historical audio, visual, haptic signals, that is $\left[\hat{A}_{t+k}, \hat{V}_{t+k}, \hat{H}_{t+k}\right] = \mathcal{P}\left(\mathbf{A}_w, \mathbf{V}_w, \mathbf{H}_w\right)$, where $\mathcal{P}(\cdot)$ is the related signal prediction function. $\hat{A}_{t+k}$, $\hat{V}_{t+k}$ and $\hat{H}_{t+k}$ are the predicted audio, visual and haptic signals. $\mathbf{A}_w = \{A_{t-w}, A_{t-w+1}, \ldots, A_t\}_{t \geq w}$, $\mathbf{V}_w = \{V_{t-w}, V_{t-w+1}, \ldots, V_t\}_{t \geq w}$, and $\mathbf{H}_w = \{H_{t-w}, H_{t-w+1}, \ldots, H_t\}_{t \geq w}$ are historical signals. Then, the predicted signals are delivered via the transmission network.

Since the received predicted signals, $\hat{A}'_{t+k}$, $\hat{V}'_{t+k}$, $\hat{H}'_{t+k}$, may be influenced by poor quality of historical signals, inadequate prediction precision, or stochastic inference throughout delivery, the receiver carries out error correction on these signals. The error correction component identifies the discrepancy between the real and the predicted signals by learning their distribution characteristics during the training procedure. This can automatically and precisely rectify the error, indicated as $\left[\tilde{V}_{t+k}, \tilde{A}_{t+k}, \tilde{H}_{t+k}\right] = \mathcal{C}\left(\hat{V}'_{t+k}, \hat{A}'_{t+k}, \hat{H}'_{t+k}\right)$, where $\mathcal{C}(\cdot)$ is the error correction function. $\tilde{A}_{t+k}$, $\tilde{V}_{t+k}$ and $\tilde{H}_{t+k}$ are the signals after error correcting. Finally, the prediction feedback component at the receiver tracks the variance between the real signals and the corrected signals, relaying this information back to the adaptive component at the sender. The adaptive component can further adjust the prediction horizon dynamically to optimize delay reduction within the bounds of prediction error tolerance.

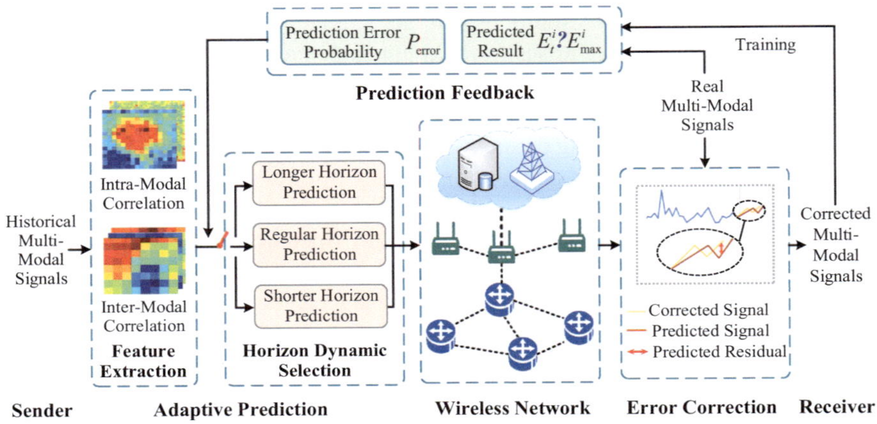

Fig. 4.11 Framework of the prediction-aware stream scheduling

On one hand, immediate prediction errors are harnessed to gauge short-term prediction uncertainties. On the other hand, it analyzes prediction error probability over an extended observation to ascertain long-term prediction uncertainty trends.

4.5.2 Cross-Modal Signal Prediction

Under the framework described above, a cross-modal signal prediction method named IGGA is designed to reduce latency. The IGGA extracts features from the perspective of intra-modal historical representation, inter-modal temporal and spatial correlation features, and then optimizes and utilizes features based on information gain. Here, we take the task of upcoming haptic signal prediction and error correction based on the historical visual and haptic signals as an example. The network structure is shown in Fig. 4.12, containing three key modules: historical feature extractor, cross-modal predictor, and error corrector.

The upcoming haptic signal prediction procedure

Specifically, different feature extractors are adopted to handle the distinct structures of multi-modal signals. For historical visual signals $\mathbf{V}_w$, an extractor $\mathcal{E}_{\mathrm{V}}$ having convolutional layers is selected to extract historical visual features, that is $F_{\mathrm{V}} = \mathcal{E}_{\mathrm{V}}(\mathbf{V}_w)$. The reason is that it has local perception capability to grasp spatial and temporal information in the visual signals. For historical haptic signals $\mathbf{H}_w$, an extractor $\mathcal{E}_{\mathrm{H}}$ with recurrent structure is selected as historical haptic feature extraction. The reason is that it has long-term memory ability to remember past status information and pass it to the next time point.

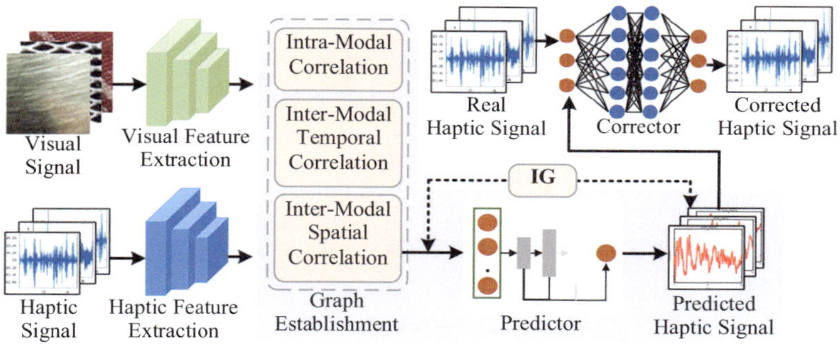

Fig. 4.12 Structure of the IGGA-based cross-modal signal prediction network

After performing feature extraction, a graph attention-based cross-modal predictor is subsequently implemented. Firstly, the graphs for visual features and haptic features are respectively denoted as $\mathcal{G}_{\mathrm{V}} = (F_{\mathrm{V}}, T_{\mathrm{V}})$ and $\mathcal{G}_{\mathrm{h}} = (F_{\mathrm{H}}, T_{\mathrm{H}})$. $F_{\mathrm{V}} = \{F_{v_1}, F_{v_2}, \ldots F_{v_w}\}$ and $F_{\mathrm{H}} = \{F_{h_1}, F_{h_2}, \ldots F_{h_w}\}$ are historical visual and haptic feature sets, which are nodes in their associated graphs. $T_{\mathrm{V}} = \{T_{v_{1,1}}, T_{v_{1,2}}, \ldots T_{v_{w,w}}\}$ and $T_{\mathrm{H}} = \{T_{h_{1,1}}, T_{h_{1,2}}, \ldots T_{h_{w,w}}\}$ are respectively temporal relationships within historical visual and haptic signals, which are adjacency matrix in their graphs.

Secondly, the intra-modal correlation is obtained from the intra-modal graph attention block, which contains a three-layer stacked graph attention layer with eight attention heads. It is noted that the number of layers and heads can be set by considering the extraction performance and complexity. Take haptic modality as a example, the intra-modal correlations can be denoted as $F^{h}_{\text{intra}} = \{F^{h_1}_{\text{intra}}, F^{h_2}_{\text{intra}}, \ldots F^{h_w}_{\text{intra}}\}$. Similarly, the inter-modal temporal and spatial correlations are extracted by the inter-modal graph attention block. For the inter-modal temporal correlation features, it can be calculated between a given haptic node and each visual graph with time-dependent nodes, represented as $F^{\mathcal{T}}_{\text{inter}} = \{F^{\mathcal{T},h_1}_{\text{inter}}, F^{\mathcal{T},h_2}_{\text{inter}}, \ldots F^{\mathcal{T},h_w}_{\text{inter}}\}$. For the inter-modal spatial correlation features, attention is considered for a given haptic node with all nodes in the visual graph, $F^{\mathcal{S}}_{\text{inter}} = \{F^{\mathcal{S},h_1}_{\text{inter}}, F^{\mathcal{S},h_2}_{\text{inter}}, \ldots F^{\mathcal{S},h_w}_{\text{inter}}\}$. The details of the extraction procedure for these three kinds of features are given in [22] and illustrated in Fig. 4.13.

Thirdly, the obtained intra-modal, inter-modal, temporal, and spatial features are fused as

$$F_f = F^{h}_{\text{intra}} \oplus F^{\mathcal{T}}_{\text{inter}} \oplus F^{\mathcal{S}}_{\text{inter}}. \tag{4.67}$$

Finally, the upcoming haptic signals $\hat{H}_{t+k}$ are predicted by inputting F_f into a fully connected layer. The prediction horizons can be determined by designing the activation fully connected layers according to the demands of the specific multi-modal services.

Concerning the error corrector, the multi-layer perception-based prediction error perception is adopted for reducing prediction error. This module ingests the real

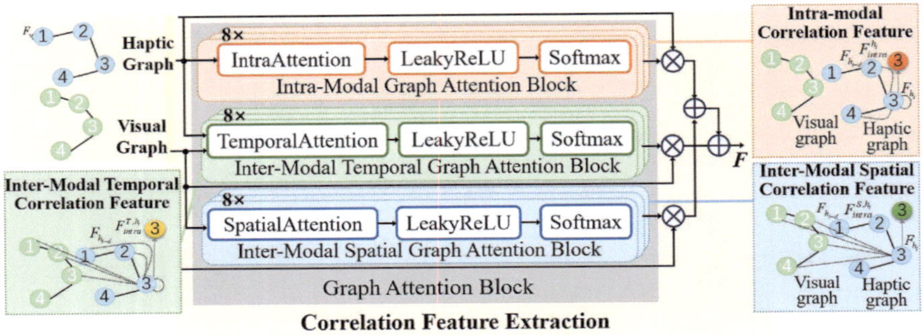

Fig. 4.13 Structure of correlation feature extraction in the IGGA

haptic signal H_{t+k} alongside the predicted haptic signal $\hat{H}_{t+k}$ to measure predictive inaccuracies and make adjustments. On one hand, the distribution features of real data are obtained through training on the real haptic signals. On the other hand, the corrector fine-tunes its parameters by assessing the disparity between the real and the predicted signals, ensuring precise rectification of prediction errors, outputting the corrected haptic signals $\widetilde{H}_{t+k}$.

Loss function design

To precisely perform the upcoming haptic signal prediction, a two-part loss function is constructed to minimize the differences between the real and the predicted signals. The first part is about the predictor optimization. Here, the smooth $L1$ loss is adopted [23], that is

$$\mathcal{L}_{pre}^{num} = \sum_{m=1}^{M} smooth_{L1}\left(\hat{H}_{t+k}(m) - H_{t+k}(m)\right). \tag{4.68}$$

Let $h = \hat{H}_{t+k}(m) - H_{t+k}(m)$, then,

$$smooth_{\mathrm{L1}}(h) = \begin{cases} 0.5\,h^2 & |h| < 1 \\ |h| - 0.5 & |h| \geq 1 \end{cases}. \tag{4.69}$$

M is the number of haptic signal channels. Moreover, to guarantee the consistency of structures, the structure-related loss is defined as

$$\mathcal{L}_{pre}^{str} = -\sum_{m=1}^{M} \mathcal{CS}\left(\hat{H}_{t+k}(m), H_{t+k}(m)\right), \tag{4.70}$$

where

$$\mathcal{CS}\left(\hat{H}_{t+k}(m), H_{t+k}(m)\right) = \frac{\hat{H}_{t+k}(m) \cdot H_{t+k}(m)}{\left\|\hat{H}_{t+k}(m)\right\| \left\|H_{t+k}(m)\right\|}. \tag{4.71}$$

Furthermore, to efficiently optimize and utilize F_f, the gap between the distribution of the predicted haptic signals $p_{\hat{H}}$ and that of the extracted features p_F should be minimized. This enables F_f to contain as much valid information as possible, enhancing the prediction accuracy,

$$\mathcal{L}_{pre}^{IG} = D_{KL}\left(p_{\hat{H}} \| p_F\right). \tag{4.72}$$

It is noted that the minimization of this loss function is derived from the maximization of information gain [24, 25], that is

$$IG(\hat{H}, F) = H(\hat{H}_{t+k}) - H(\hat{H}_{t+k} \mid F_f). \tag{4.73}$$

$H(\hat{H}_{t+k})$ denotes the entropy of the predicted signal $\hat{H}_{t+k}$, and $H(\hat{H}_{t+k} \mid F_f)$ denotes the conditional entropy of $\hat{H}_{t+k}$ given F_f. Under F_f, a larger IG means that the uncertainty in $\hat{H}_{t+k}$ is relieved more. In other words, F_f is more helpful in enhancing prediction precision.

From the above considerations, the loss function for the first part is denoted as,

$$\mathcal{L}_{pre} = \alpha_1 \mathcal{L}_{pre}^{num} + \alpha_2 \mathcal{L}_{pre}^{str} + \alpha_3 \mathcal{L}_{pre}^{IG}, \tag{4.74}$$

where $\alpha_1, \alpha_2, \alpha_3 \in [0, 1]$ are the associated weights. The second part is about joint optimization of the predictor and error corrector. $L1$ loss is used for learning the characteristics of the real haptic signals,

$$\mathcal{L}_{cor}^{real} = \sum_{m=1}^{M} \left| H_{t+k}(m) - \mathcal{C}\left(H_{t+k}(m)\right) \right|, \tag{4.75}$$

where $\mathcal{C}(\cdot)$ is the error corrector. Similar to the loss functions in the first part, smooth $L1$ loss and negative correlation losses are constructed for describing the gap between $\tilde{H}_{t+k}$ and H_{t+k},

$$\mathcal{L}_{err}^{num} = \sum_{m=1}^{M} smooth_{L1}\left(\tilde{H}_{t+k}(m) - H_{t+k}(m)\right), \tag{4.76}$$

$$\mathcal{L}_{err}^{str} = -\sum_{m=1}^{M} \mathcal{CS}\left(\tilde{H}_{t+k}(m), H_{t+k}(m)\right). \tag{4.77}$$

Therefore, the loss of the second part is represented as

$$\mathcal{L}_{err} = \gamma_1 \mathcal{L}_{cor}^{real} + \gamma_2 \mathcal{L}_{err}^{num} + \gamma_3 \mathcal{L}_{err}^{str}, \quad (4.78)$$

where $\gamma_1, \gamma_2, \gamma_3 \in [0, 1]$ are the associated weights.

4.5.3 Latency and Prediction Error Analysis

In order to analyze the advantages of the designed prediction-aware stream scheduling scheme, latency and prediction error analysis should be employed. Here, packet length is assumed to be fixed and larger than the size of the predicted signal. The bandwidth is larger than the packet delivery rate. Network configurations in both the prediction-aware and the non-prediction-aware schemes are identical. The perceived delay $\hat{D}$ is denoted as the duration from when the signal is generated at the sender to when it is perceived by the receiver. The prediction error probability P_{error} is denoted as the probability that the error between the corrected signal and the real signal for at least one modality surpasses the maximum allowable prediction error for that modality. Based on this definition, the prediction error probability for simultaneously predicting the upcoming audio, visual, and haptic signals is

$$P_{\text{error}} = 1 - \Pr\left\{E_t^A \leq E_{\text{max}}^A, E_t^V \leq E_{\text{max}}^V, E_t^H \leq E_{\text{max}}^H\right\}, \quad (4.79)$$

where $E_{\text{max}}^i, i \in \{A, V, H\}$ are the corresponding maximum allowable prediction errors at time t.

For *latency analysis*, the delay about the non-prediction-aware and the prediction-aware schemes is provided in Fig. 4.14. In the non-prediction-aware scheme, the P-data begins transmission from the sender at t_3 and arrives at the receiver at t_5, resulting in an end-to-end (E2E) delay represented as $(t_5 - t_3)$. In the prediction-aware scheme, assuming a historical window w of 5 and a prediction horizon k of 4, the upcoming P-data is forecasted 4 time instances ahead at t_1. The prediction is finalized after a duration of $(t_2 - t_1)$. Following this, the predicted P-data is transmitted at t_2 and received at t_4, with the E2E delay expressed as $(t_4 - t_2)$. It is evident that the inclusion of the prediction module effectively reduces the delay by $(t_5 - t_4)$ when compared to the scheme without prediction. In general, the absolute delay $\tilde{D}$ is denoted as the sum of the E2E delay D and prediction delay T_p, $\tilde{D} = D + T_\text{p}$. For each time slot T_s and prediction horizon k, the perceived delay is

$$\hat{D} = \tilde{D} - kT_\text{s} = D + T_\text{p} - kT_\text{s}. \quad (4.80)$$

Based on the above definitions, a broader prediction horizon results in a reduced $\hat{D}$. Specifically, if the prediction horizon is extended while keeping other parameters constant, the receiver gets the predicted data earlier, thereby decreasing $\hat{D}$. Under the assumption of unconstrained computation and communication resources, D could be compensated by increasing k, leading to zero-latency transmission in the theoretical aspect [2]. Moreover, even though the prediction process might increase perceived

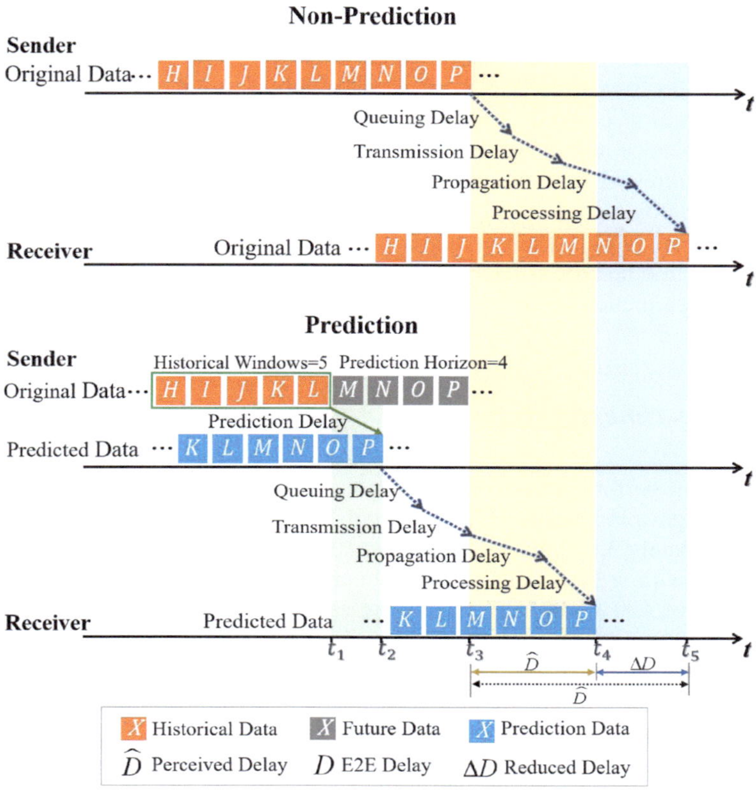

Fig. 4.14 Delay analysis of the scheme without prediction versus the proposed scheme with prediction [22]

delay, this increase becomes negligible if the processing time of the method is sufficiently short or if the k is sufficiently large. Consequently, the receiver can obtain the desired signals from the sender in advance, enabling it to perceive the sender's state in a timely manner and thus minimize $\hat{D}$.

For *prediction error analysis*, the prediction module inevitably generates prediction error, which should be reduced. The parameter k affects the ability of this scheme for capturing upcoming trends [26]. In fact, short-term prediction usually leads to high prediction precision, but it cannot grasp long-term trends and results in the increase of $\hat{D}$. From this analysis, by performing prediction of the upcoming signals and error correction, $\hat{D}$ can be minimized. Take the M-channel haptic signals as an example, after implementing optimization about Eq. (4.78), the prediction error probability can be represented as

$$P_{\text{error}}^{H} = 1 - \Pr\left\{E_t^H(1) \leq E_{\max}^H, \ldots, E_t^H(M) \leq E_{\max}^H\right\}. \tag{4.81}$$

Moreover, the threshold of prediction error probability should be set based on the prediction precision demands of the specific multi-modal services. If the prediction error probability exceeds the preset threshold, it implies that the prediction precision is insufficient to meet the requirements, necessitating a reduction in the prediction horizon to enhance accuracy. Conversely, when the probability of prediction error is below the threshold and the prediction error is within the acceptable maximum for each modality signal, the prediction precision is deemed to consistently meet the requirements and is likely to significantly decrease perceived latency; thus, a longer prediction horizon is adopted. In other cases, based on additional feedback, the standard prediction horizon is preserved.

4.5.4 Experimental Result and Analysis

To evaluate the performance of this prediction-aware stream scheduling scheme, the multi-modal grasping dataset [27] is selected. This dataset is generated by conducting approximately 3,000 grasping trials on 10 different objects using robotic arms and hands, yielding both visual and haptic signals. The visual signal, captured at a resolution of 640×480, has a sampling rate of 18 Hz, while the haptic signal is obtained from a 16-channel piezoresistive sensor with a sampling frequency of 40 Hz in mN. Then, the cross-modal signal prediction in Sect. 4.5.2 is trained on this dataset, which predicts the upcoming haptic signals based on historical visual and haptic signals. The competing approaches include LSTM [28], GMAN [29], and Transformer [30]. When concerning evaluation metrics, the mean absolute error (MAE) is used to measure the absolute difference between the real signal and the predicted haptic signal. The prediction accuracy (ACC) assesses the performance by quantifying the absolute discrepancy between the real and the predicted haptic signals, effectively considering both positive and negative variations.

$$ACC = \begin{cases} 100\% - \frac{100\%}{M} \sum_{m=1}^{M} \left| \frac{H_{t+k}(m) - \tilde{H}_{t+k}(m)}{H_{t+k}(m)} \right| & 0 < \frac{\tilde{H}_{t+k}(m)}{H_{t+k}(m)} < 2 \\ 0 & \text{otherwise.} \end{cases} \tag{4.82}$$

To maintain experimental consistency, the historical window is standardized to 12, and the prediction horizons are varied across all approaches. The result about MAE is presented in Table 4.1. Notably, the MAE for all prediction methods increases with an extended prediction horizon. The proposed prediction scheme outperforms the baseline. Among the various methods, LSTM exhibits the highest MAE values, attributed to its reliance on a gating mechanism for state updates. This makes it prone to outliers and noises. Additionally, while GMAN, which is graph-based, can capture node relationships and features, it lacks sensitivity to data sparsity, leading to reduced accuracy in predicting haptic signals with discrete characteristics. Although the Transformer is adept at capturing long-range dependencies, its lack of robust

Table 4.1 MAE performance comparison of different approaches in different prediction horizons

Approach	$k = 1$	$k = 2$	$k = 3$	$k = 4$	$k = 5$	Average
LSTM	3.241	6.523	7.321	7.611	8.128	6.565
Transformer	1.798	4.895	5.652	6.023	6.784	5.031
GMAN	2.359	4.892	5.354	6.396	6.612	5.123
Ours	**1.672**	**4.340**	**5.023**	**5.901**	**6.002**	**4.588**

positional information processing leads to inferior haptic signal predictions compared to the proposed approach.

Next, a qualitative comparison between the baseline and the proposed prediction approach is also performed. Figure 4.15 illustrates the predicted haptic signals for each approach with a historical window of 12 and a prediction horizon of 2. The proposed approach surpasses the baseline in predicting more plausible and detailed haptic signals. Specifically, while the baseline methods can capture the general trend of the haptic signal, they fail to accurately forecast it during abrupt emergences or significant changes. Furthermore, the baseline schemes exhibit substantial deviations in predicting the amplitude of haptic signals. This underscores the superiority of the proposed approach in terms of effectiveness.

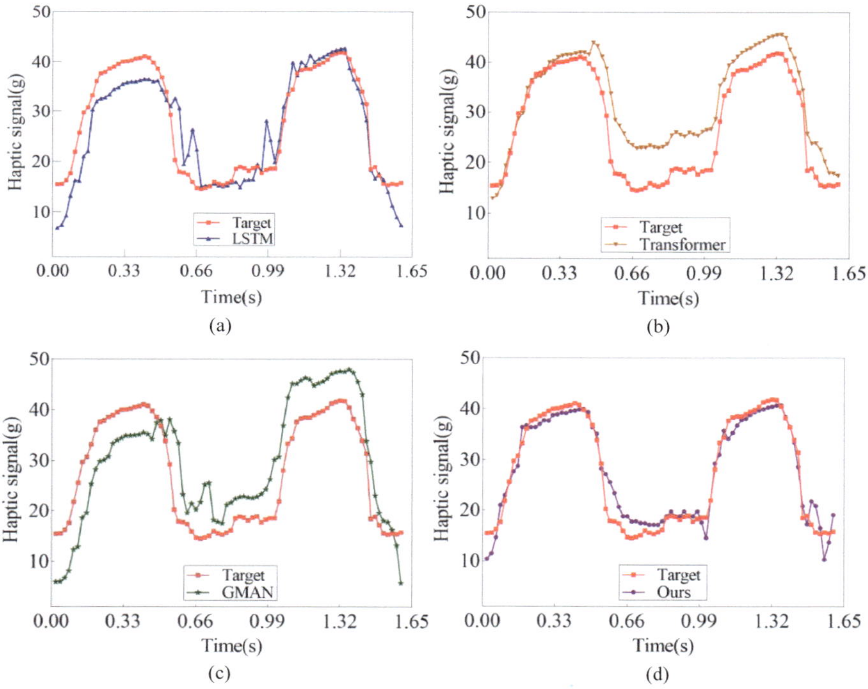

Fig. 4.15 Comparison of prediction results of the proposed scheme and the baseline on the multi-modal grasping dataset

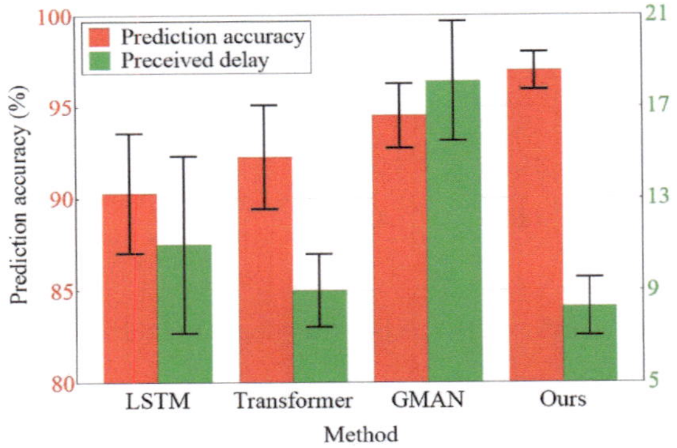

Fig. 4.16 The performance of IGGA and baselines about prediction accuracy and perceived delay

To further assess the practical effectiveness, the prediction precision and perceived latency of the baseline and the proposed scheme are compared. A transmission rate of 1 Mbps over a distance of 1000 km are specified. Given that single-mode fiber optics is utilized for transmission, the propagation speed is roughly 2×10^8 m/s. The communication channel is characterized by random, additive white Gaussian noise under optimal conditions. Moreover, a consistent initial prediction horizon of 2 and a historical window size of 10 are adopted. Data collection commences from the initial data point at $t = 0$, with intervals determined by the historical window. Future signal prediction commences once the accumulated signal meets the historical window requirements. Furthermore, a prediction error probability threshold of 0.1 and the maximum acceptable error for haptic signal prediction of 5 are set. From Fig. 4.16, it is notable that the proposed scheduling scheme possesses a certain edge in prediction accuracy, enabling it to extend the prediction horizon effectively to counterbalance the processing time, thereby resulting in a lower perceived delay compared to the baseline schemes. To sum up, the proposed scheme accomplishes precise prediction and reduced latency, which is suitable for cross-modal transmission.

4.6 Dynamic Transmission Mode Selection

To handle the heterogeneous transmission requirements of haptic and audio-visual signals, separate transmission seems to be an available solution. However, this will consume more communication resources. When several terminal devices simultaneously access the communication system, a sharp drop in performance may happen, leading to increased power consumption. Different from existing solutions, this section considers efficient multi-modal streaming from the perspective of transmission mode selection [31]. Firstly, for haptic-dominant, audio-visual dominant, and equally important multi-modal services, the corresponding transmission strategies are designed. Then, by estimating the resource requirements of different modality streams, suitable transmission modes are selected, and resources such as bandwidth, power, and relay nodes are also allocated.

4.6.1 Representative Transmission Strategy Construction

In order to fulfill the transmission requirements of heterogeneous streams, three transmission strategies are designed, shown in Fig. 4.17. Here, different modality streams are delivered over distinct links and enjoy separate buffers. Users are categorized into single-modal service user (SU) and multi-modal service user (MU). The former employs communication by using cellular mode, while the latter realizes communication by these three transmission strategies described in the following.

Strategy A is designed for audio-visual-dominant multi-modal services. In these services, channel capacity is required for audio-visual streams, while the latency demand for haptic streams can be appropriately relaxed. Thus, audio-visual streams can be delivered via the D2D mode or the cellular mode, needing to allocate more bandwidth. Haptic streams can be delivered by the multiplexing mode. Each MU can utilize the uplink resources of an SU for communication. The transmission rate of audio-visual in cellular or D2D mode is given by the expression:

$$R^a_{V_{CEL}}(j) = min\left\{B^a_V(j)\log_2\left(1+\frac{P^a_V(j)G^a_{TB}(j)}{n_0 B^a_V(j)}\right), B^a_V(j)\log_2\left(1+\frac{P_{BR}G^a_{BR}(j)}{n_0 B^a_V(j)}\right)\right\}, \tag{4.83}$$

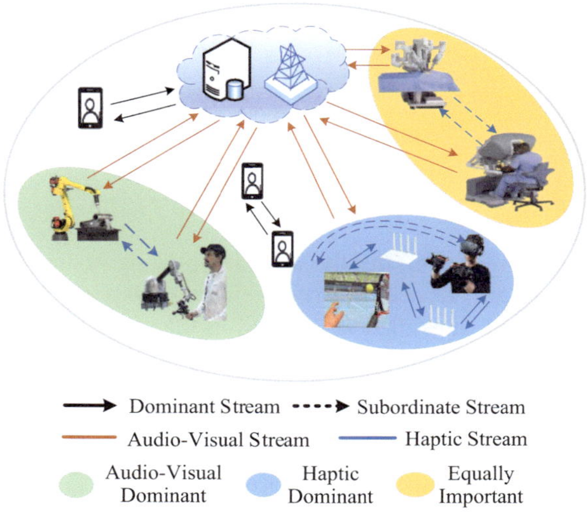

Fig. 4.17 Transmission strategies for multi-modal services

$$R^a_{V_{DIR}}(j) = B^a_V(j)\log_2\left(1+\frac{P^a_V(j)G^a_{TR}(j)}{n_0 B^a_V(j)}\right). \tag{4.84}$$

To ensure low latency for haptic streams, the way of transmitting a small number of packets in short bursts is usually utilized. In such scenarios, the additive noise and distortion caused by the channel are not easily mitigated, making it challenging for the Shannon formula to precisely determine the channel capacity. Instead, the finite block length channel coding is employed, with a rate that is roughly as described in [32].

$$\begin{aligned} R^a_H(j) = B^a_H(j)\log_2\left(1+\frac{P^a_H(j)G^a_{TR}(j)}{P_{SU}G^a_{SR}(j)+n_0B^a_H(j)}\right) \\ -\sqrt{\frac{v^a_H(j)}{L^a_H(j)}}Q^{-1}\left(\varepsilon^c\right)+O\left(\frac{\log L^a_H(j)}{L^a_H(j)}\right). \end{aligned} \tag{4.85}$$

Here, a represents that this user is employing strategy A. B is channel bandwidth, and P is transmit power. V and H respectively denote audio-visual and haptic modalities. n_0 is additive white Gaussian noise and G is channel gain. TB, BR, TR, and SR, respectively, represent links from the sender to the base station, from the base station to the receiver, from the sender to the receiver, and from the multiplexed user to the receiver. v is channel dispersion, L is the mean packet length follows an exponential distribution, ε^c is packet error probability, and Q is Gaussian Q-function.

Strategy B is designed for haptic-dominant multi-modal services. In these services, low latency is required for haptic streams. Thus, the multiplexing mode is assigned for audio-visual streaming, that is

$$R^b_V(k) = B^b_V(k)\log_2\left(1+\frac{P^b_V(k)G^b_{TR}(k)}{P_{SU}G^b_{SR}(k)+n_0B^b_V(k)}\right). \tag{4.86}$$

Haptic streams are delivered by the D2D or relay mode. For the D2D mode, the end-to-end delay of haptic streaming is

$$T^b_{H_{DIR}}(k) = T^b_{sendout}(k) + T^b_{transmit}(k) + T^b_{processing}(k) + T^b_{queuing}(k). \tag{4.87}$$

Here, b represents that this user is employing strategy B. $T_{sendout}$, $T_{transmit}$, $T_{processing}$, and $T_{queuing}$ respectively represent transmission delay, propagation delay, processing delay, and queuing delay. $T_{sendout}$ is the duration from sending the first bit to the last bit of the data block [33],

$$T^b_{sendout}(k) = \frac{L^b_H(k)}{R^b_{H_{DIR}}(k)}. \tag{4.88}$$

In Eq. (4.88), $R^b_{H_{DIR}}(k)$ can be obtained by finite block length coding rate [32]. As haptic streaming happens on orthogonal channels, interference among channels can be ignored:

$$R^b_{H_{DIR}}(k) = B^b_H(k)\log_2\left(1 + \frac{P^b_H(k)G^b_{TR}(k)}{n_0 B^b_H(k)}\right) - \sqrt{\frac{v^b_H(k)}{L^b_H(k)}}Q^{-1}\left(\varepsilon^c\right) + O\left(\frac{\log L^b_H(k)}{L^b_H(k)}\right). \tag{4.89}$$

Let $T_{queuing}$ represent the duration during which packets are held in a queue awaiting processing. It is posited that the arrival rate of haptic packets follows a Poisson distribution with an average rate of λ_k. The average service rate is denoted as μ_k. This aligns with the $M/M/1$ queuing model. Given that the packet loss rate is negligible, λ_k can be formulated as $N^b_H(k)/L^b_H(k)$, where $N^b_H(k)$ signifies the incoming bit rate of the data. Meanwhile, μ_k is equivalent to the proportion of channel capacity to the packet length, which can be expressed as $\mu_k = R^b_{H_{DIR}}(k)/L^b_H(k)$. Thus,

$$T^b_{queuing}(k) = \frac{L^b_H(k)}{R^b_{H_{DIR}}(k)}\frac{1}{\frac{R^b_{H_{DIR}}(k)}{N^b_H(k)} - 1}. \tag{4.90}$$

In certain scenarios, the considerable distance separating the sender and the receiver leads to a diminished transmission rate and an inability to meet the desired latency. To curtail the waiting time in queues and alleviate the strain on the device's data processing capabilities, it might be essential to opt for a relay mode for certain haptic data streams. The haptic delay, when employing relay nodes, can be calculated as the cumulative delay of all the nodes that are directly linked along the communication route.

When the distance between sender and receiver is far, this may lead to a decline in transmission rate and unexpected latency. To curtail the waiting time in queues and alleviate the strain on the device's data processing capabilities, relay mode is more recommended for haptic streaming. Under these circumstances, haptic delay can be calculated as the cumulative delay of all the nodes that are directly linked along the communication route,

$$T^b_{H_{REL}}(k) = \sum_{e,f\in cell} \rho_{e,f}(k)\left[\frac{L^b_H(k)}{R^{e,f}_{H_{DIR}}(k) - N^b_H(k)}\right], \tag{4.91}$$

where e, f are respectively the sender and receiver in cellular. $\rho_{e,f}$ is node selection factor. When $\rho_{e,f} = 1$, e, f are two nodes in the haptic stream transmission path; when $\rho_{e,f} = 0$, they are not in the path. The maximum number of hops is determined by the distance d between user pairs. Moreover, the achievable rate of haptic stream $R^b_{H_{REL}}(k)$ can be expressed as the minimum value in $R^{e,f}_{H_{DIR}}(k)$.

Strategy C is designed for equally important multi-modal services. In these services, low latency is required for haptic streams, and large bandwidth is demanded for audio-visual streams. For audio-visual streams, one can choose either D2D mode or cellular mode, similar to Strategy A, with the option to allocate additional channel bandwidth. For haptic streams, one can select either D2D mode or relay mode, as in Strategy B, to identify a path that offers low latency.

4.6.2 Bandwidth and Delay Threshold Estimation

The aim of these three representative strategies is to ensure the delivery of high-quality multi-modal services. The key to achieving this objective lies in the precise estimation of the transmission requirements, as described in the following.

For audio-visual streams, they exhibit characteristics of self-similarity. The self-similarity refers to their possession of similar features in the frequency domain or time domain over a certain period. Consider this, a long-range dependent traffic prediction model [34] is employed here to achieve constrained computation for audio-visual streams. On the basis of analyzing the statistical characteristics of network traffic, this model efficiently predicts the minimum bandwidth required to ensure the QoS of audio-visual streaming with low time complexity. Specifically, audio-visual traffic at the receiver during the period $0 \sim t$ is assumed to conform to fractional Brownian motion, $D(t) = mt + \sqrt{um}F_t$, where F_t is the standard fractional Brownian motion. m is the average arrival traffic of audio-visual streams and u is the associated variance,

$$m = \frac{p_w p_l p_b f_v}{c_v} + \frac{f_a s_a n_a}{c_a}. \tag{4.92}$$

p_w and p_l are the horizontal and vertical resolutions of the visual signal. p_b is the size of sites per pixel. f_v is the average number of frames per second of the visual signal. f_a is the average sampling frequency of the audio signal. s_a is the sampling accuracy of audio. n_a is the number of channels of the device. c_v and c_a are the average compression rates of visual and audio signals. When the transmission rate of the audio-visual stream $R > m$, the probability that the queue length is greater than y can be expressed as $\varepsilon = P\{Y(t) > y\}$. Therefore, the estimated bandwidth is

$$R_{a\&v} = m + \left[H_u^{H_u}(1 - H_u)^{1-H_u}\sqrt{-2\ln\varepsilon}\right]^{\frac{1}{H_u}} u^{\frac{1}{2H_u}} y^{\frac{H_u-1}{H_u}} m^{\frac{1}{2H_u}}, \tag{4.93}$$

where H_u is Hurst exponent.

For haptic streams, research indicates that when the time delay of haptic signals is controlled within a certain range, haptic feedback does not affect human discriminative perception [34]. In other words, under these circumstances, users can achieve a sensation nearly identical to local operation when manipulating remote devices. To ensure the timely arrival of haptic streams and the rational use of resources, it is necessary to set a delay threshold for haptic signals. Mode selection and resource allocation must be based on the premise that the delay of haptic streams does not exceed the set threshold.

The haptic delay threshold here is based on the transparency theory. Transparency specifically reflects the accuracy with which a remote environment is presented through a haptic network. Ideally, user's sensation at the command sender should be consistent with the sensation at the receiver. Transparency is of significant importance for services related to haptic data streams because it can ensure the realism of the remote operating environment. A typical objective indicator of transparency is the consistency of impedance between sender and receiver. When user's impedance matches the environment's impedance, transparency is maximized [35]. Specifically, the environmental impedance of the sender can be approximated as [36]:

$$Z_T(s) \approx g_w \frac{Z_R(s) + g_w T_H s}{g_w + T_H s Z_R(s)}, \tag{4.94}$$

where $s = j\omega$. j is the square root of -1, ω is the angular frequency of the haptic information. $Z_R = k_r/s$ is the environmental impedance of receiver, where k_r is stiffness of receiver environment. g_w represents wave impedance. T_H represents delay of haptic signal. Substituted these items into Eq. (4.94),

$$Z_T(s) = \frac{k_t}{s}\left(1 + \frac{g_w T_H s^2}{k_r}\right), \tag{4.95}$$

where k_t represents the stiffness perceived by the sender,

$$k_t = \frac{g_w k_r}{g_w + T_H k_r}. \tag{4.96}$$

In order to ensure that the haptic difference between sender and receiver is not perceived by humans,

$$k_t > (1 - \mathrm{JND}_k)\, k_r, \tag{4.97}$$

where $\mathrm{JND}_k \in (0, 1)$ denotes the difference perception threshold of hardness. Substitute this item into Eq. (4.96), and the delay threshold of haptic streaming can be obtained

$$T_{th} = \frac{\mathrm{JND}_k}{1 - \mathrm{JND}_k} \frac{g_w}{k_r}. \tag{4.98}$$

It is noted that when the stiffness parameters of the environment cannot be easily obtained, the delay threshold can be directly set, i.e., $1ms$ about the most rigid delay.

4.6.3 Joint Transmission Mode Selection and Resource Allocation

After constructing the transmission strategies and constraints for different kinds of multi-modal services, the specific modes for MU should be selected. For audio-visual-dominant services, the audio-visual rate is $R_V^a(j) = \sum_{m \in M1} \alpha_j^m R_{V_m}^a(j)$, $\alpha_j^m = \{0, 1\}$.$\alpha_j^m = 1$ indicates that user j selects mode m, $M1 = \{CEL, DIR\}$. CEL, DIR respectively denote the cellular mode and the D2D mode. For haptic-dominant services, haptic delay is $T_H^b(k) = \sum_{m \in M2} \beta_k^m T_{H_m}^b(k)$, $\beta_k^m = \{0, 1\}$. $\beta_k^m = 1$ indicates that user k selects mode m, $M2 = \{DIR, REL\}$. REL denotes the relay mode. The haptic rate is $R_H^b(k) = \sum_{m \in M2} \beta_k^m R_{H_m}^b(k)$. For equally important multi-modal services, audio-visual rate is $R_V^c(l) = \sum_{m \in M1} \gamma_l^m R_{V_m}^c(l)$. Haptic delay is $T_H^c(l) = \sum_{m \in M2} \delta_l^m T_{H_m}^c(l)$, and haptic rate is $R_H^c(l) = \sum_{m \in M2} \delta_l^m R_{H_m}^c(l)$. Roles of γ_l^m and δ_l^m are the same as α_j^m and β_k^m. Moreover, resources should be carefully allocated to the relay nodes and terminal users, which contain channel bandwidth, transmission power, and the devices in the cell. Based on this, the multi-objective optimization problem can be established.

$$\max \left\{ \sum_{j=1}^{N1} R_V^a(j) + \sum_{l=1}^{N3} R_V^c(l) \right\}, \tag{4.99}$$

$$\min \left\{ \sum_{k=1}^{N2} T_H^b(k) + \sum_{l=1}^{N3} T_H^c(l) \right\}, \tag{4.100}$$

$$\max \left\{ \sum_{j=1}^{N1} \left[R_V^a(j) + R_H^a(j) \right] + \sum_{k=1}^{N2} \left[R_V^b(k) + R_H^b(k) \right] + \sum_{l=1}^{N3} \left[R_V^c(l) + R_H^c(l) \right] \right\}, \tag{4.101}$$

$$\max \left\{ R_{\text{total}} / \left[\sum_{j=1}^{N1} \left(P_V^a(j) + P_H^a(j) \right) + \sum_{k=1}^{N2} \left(P_V^b(k) + P_H^b(k) \right) + \sum_{l=1}^{N3} \left(P_V^c(l) + P_H^c(l) \right) \right] \right\}, \tag{4.102}$$

$$\text{s.t. } P_{\min} \le P_V^a(j), P_H^a(j), P_V^b(k), P_H^b(k), P_V^c(l), P_H^c(l) \le P_{\max}, \forall j, k, l, \tag{4.103}$$

$$\sum_{j=1}^{N1} B_V^a(j) + \sum_{l=1}^{N3} B_V^c(l) \le B_{\max}, \forall j, l, \tag{4.104}$$

$$R_V^a(j) \ge R_{a-v}^a(j), R_V^c(l) \ge R_{a-v}^c(l), \forall j, l, \tag{4.105}$$

$$T_H^b(k) \le T_{th}^b(k), T_H^c(l) \le T_{th}^c(l), \forall k, l. \tag{4.106}$$

Equation (4.99) is about maximizing audio-visual rates for audio-visual-dominant services and equally important services. Equation (4.100) is about minimizing haptic delays for haptic-dominant services and equally important multi-modal services. Equations (4.101) and (4.102) are about maximizing throughput and energy utilization, respectively. $N1$, $N2$ and $N3$ respectively denote the number of user pairs in strategy A, strategy B, and strategy C. R and T are respectively the channel capacity and delay. R_{total} represents the throughput of MU. Equation (4.104) represents the transmission power constraint with the maximum $P_{\max}$ and minimum $P_{\min}$. Equation (4.103) is about channel bandwidth constraint, with the total available spectrum as $B_{\max}$. Equations (4.105) and (4.106) respectively represent the constraints of audio-visual and haptic streams.

Finally, for optimizing the problem, a joint mode selection and resource allocation algorithm is designed, described in Algorithm 4.2. To be more specific, to minimize the capacity loss, the device in the cell most distant from the current user should be identified for reuse. The NSGA-III algorithm [37] is employed to produce a non-dominated solution set $\widetilde{P}_t$ through a limited number of iterations. If the channel capacity fails to meet the requirements for audio-visual streaming after these iterations, the stream is further divided, and the excess traffic is sent via a multiplexing link, resulting in the formation of a multiplexed entity. If the delay does not meet the haptic stream's requirements, the network is treated as a weighted graph, with the weights indicating haptic latency. The path of minimum delay is determined using a dynamic programming approach. In conclusion, the algorithm yields $\widetilde{P}_t$, the multiplexed entity, and the path of minimum delay.

Algorithm 4.2 Joint mode selection and resource allocation algorithm [31]

Require: $R_{a-v}(i) \leftarrow m(i), H_u(i), u(i), y(i), \varepsilon(i)$; $T_{th}(i) \leftarrow \text{JND}_k(i), g_w(i), k_r(i)$, max number of hop $hop(i) \leftarrow d(i)$

1: Initialization: Define reference points and initialize population P_t containing modes, relay nodes, transmit power and bandwidth for MU pairs.
2: **Repeat**
3: $P_t \leftarrow$ NSGA-III algorithm
4: $t = t + 1$
5: **Until** obtaining a set of non-dominated solution $\widetilde{P}_t$
6: **If** $R_V(i) < R_{a-v}(i)$ **then**
7: Overflow traffic is transmitted over multiplexed links, and select the multiplexed object;
8: **If** $T_H(i) > T_{th}(i)$ **then**
9: Calculate and record the minimum delay path using dynamic programming algorithm.

Output: $\widetilde{P}_t$, multiplexed object, minimum delay path

4.6.4 Experimental Result and Analysis

The simulated setting consists of a hexagonal cell that spans 500 m in radius. Positioned at the cell's core, the base station stands 25 m tall and has a peak transmission power of 46 dbm. Each resource block occupies a bandwidth of 180 KHz, and the ambient Gaussian noise level is −174 dbm/Hz. The total bandwidth amounts to 26 MHz. Within this setup, there are 70 SUs and 25 pairs of MUs, each capable of a maximum transmission power of 27 dbm. The transmission power for a single stream ranges from a minimum of 20 dbm to a maximum of 24 dbm. Moreover, the WINNER-II path loss model is utilized in this experiment [38].

The performance of the simulation is benchmarked against the other three approaches: D2D [38], cellular [39], and the constraint-based mode selection algorithm (CB-MSA) [40]. For these approaches, resource blocks are evenly distributed across the spectrum. Users are assigned an equal number of resource blocks, which

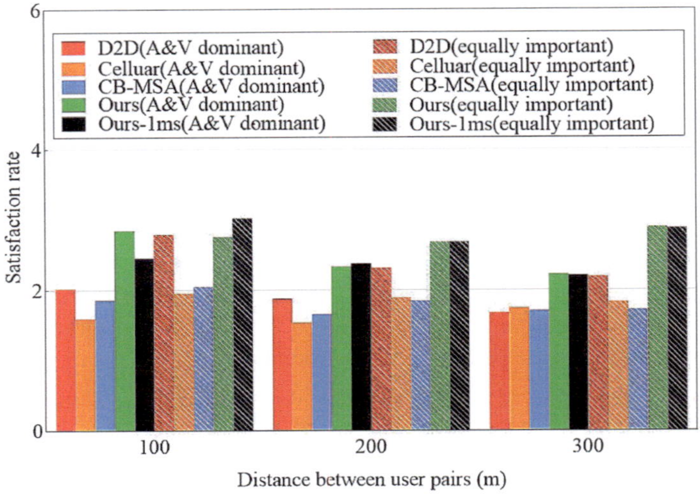

Fig. 4.18 Satisfaction rate of audio-visual-dominant and equally important multi-modal services

are selected at random within the system's bandwidth. Furthermore, the performance of the proposed algorithm under a $1ms$ delay threshold is also demonstrated.

Figure 4.18 illustrates the satisfaction rate for audio-visual-dominant and equally important multi-modal services. Satisfaction rate is calculated as the ratio of the audio-visual link rate to the imposed constraint. As audio, visual, and haptic signals are delivered according to different modes, the transmission rate of audio-visual streams is unaffected by haptic streams. The proposed algorithm exhibits a comparable satisfaction rate in both scenarios of delay threshold prediction and a $1ms$ delay threshold, surpassing that of the other competing algorithms.

Figure 4.19 illustrates the average delay of haptic signals for haptic-dominant and equally important multi-modal services. It is evident that the average delay of the proposed algorithm, in both delay threshold prediction and at an 1 ms delay threshold, is considerably lower than the competing algorithms. This low delay is achieved for two main reasons. On one hand, the queuing advantage from the heterogeneous stream splitting transmission, where haptic packets are queued without the need to wait for audio-visual packets, significantly reduces queuing delay. On the other hand, a more effective transmission path can be planned, leading to a decrease in latency.

Figures 4.20 and 4.21 respectively present the throughput and the energy utilization rates. The proposed algorithm that relies on delay threshold prediction demonstrates the highest performance. The version of the proposed algorithm set at a 1 ms delay threshold is marginally less effective than the former but still notably outperforms the other comparative algorithms. This superior performance is attributed to the model's allocation of two separate links for the transmission of heterogeneous signals, leading to a multichannel benefit. Additionally, the integration of power control allows the algorithm to dynamically adjust transmission power to meet service

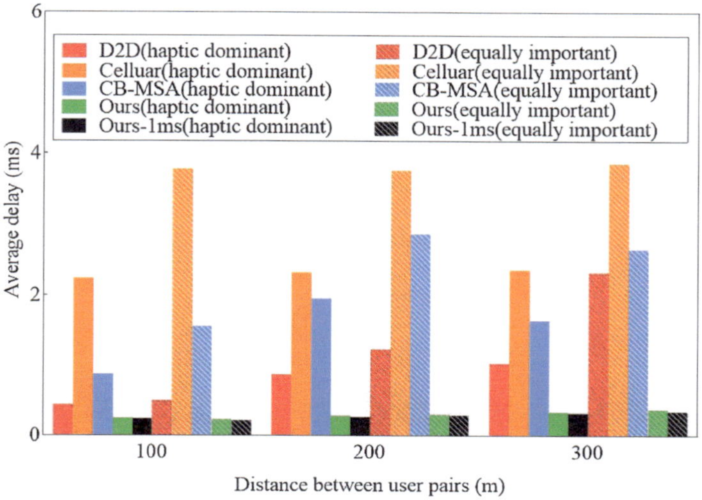

Fig. 4.19 Average haptic delay of haptic-dominant and equally important multi-modal services

Fig. 4.20 Throughput of multi-modal services

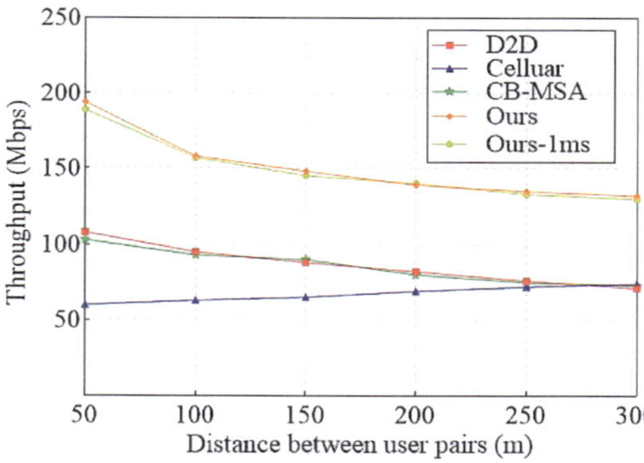

demands. The NSGA-III algorithm's robust search capabilities further enhance the overall performance.

4.7 Edge Intelligence-Empowered Transmission Decision

From the discussions in the previous sections, stringent latency for haptic streaming and fierce resource competition among modalities are the two essential issues to pay attention to. Besides, from the perspective of applications, two other factors affecting cross-modal transmission also need to be carefully considered. On one hand,

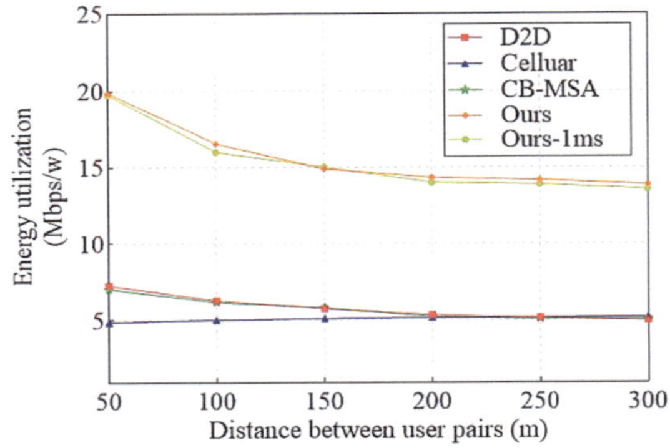

Fig. 4.21 Energy utilization of multi-modal services

due to variable wireless channel states and high mobility of terminals, the network environment may evolve over time throughout the cross-modal transmission process. On the other hand, due to diverse service contexts, transmission requirements such as low latency, high reliability, and high throughput should also be dynamically changed and adjusted in time. Edge computing has been considered as a promising paradigm to support large-scale multimedia streaming. Although several works have emerged about caching, computation, and communication framework construction and specific scheme design in edge computing, they mainly focus on single-modal streaming with limited technical indicators [41, 42]. Moreover, AI is recognized as a significant tool deployed in edge servers to further contribute "wisdom" for applications. Clearly, the integration of edge computing with AI, known as edge intelligence [43], can optimize the utilization of networking resources for multi-modal streaming from an intelligent standpoint. Therefore, this section proposes an edge intelligence-empowered cross-modal streaming framework, considering communication, caching, computation, and control capabilities (4C) [44]. Under this framework, the optimal decision is employed based on the attention-based broad reinforcement learning, which can reasonably allocate networking resources.

4.7.1 Joint Communication, Caching, Computation and Control

Here, the edge intelligence-empowered cross-modal streaming framework contains secure wireless communication, efficient content caching, collaborative computation offloading, and adaptive optimization control. Firstly, taking into account user privacy and data security, federated learning and blockchain techniques are employed for device-to-edge and D2D communications to ensure secure wireless communication. Secondly, to achieve more efficient content caching from sparse data, clustering algorithms are implemented, and transfer learning is integrated into distributed collaborative filtering for rapid estimation of content popularity. Thirdly, unlike traditional approaches that offload tasks entirely to user equipment (UE) or the edge, computing tasks can be completed through the collaboration of UEs and edge servers to enhance energy efficiency. Fourthly, to make the entire framework adaptable to dynamic network conditions, diverse service scenarios, and heterogeneous streams, a control mechanism is developed to address the joint optimization problem of secure communication, efficient caching, and collaborative computation. The details are described in the following.

Secure wireless communication

To facilitate the delivery of content or the offloading of tasks, there will be an associated communication expense for the UE, edge union, and central cloud. Here, four communication scenarios can be considered: D2D, device-to-edge communications, edge-to-edge communications, and edge-to-cloud communications. Given that edge-to-edge and edge-to-cloud connections are supported by fiber-optic links, the discussion is limited to D2D and device-to-edge communications, which are interconnected via wireless channels. For UEs with limited resources, the data rate is determined in accordance with Shannon's Theory [42].

To guarantee data security and prevent privacy breaches during multi-modal streaming, the blockchain-enhanced D2D communication and federated learning-based device-to-edge communication [45] can be performed. For the former, blockchain can enable decentralized, distributed, and synchronized identification. When a target UE requires assistance from others in transmitting or sharing content, it may initiate by broadcasting a request along with transaction details. Subsequently, the other UEs decide whether to respond based on the anticipated gains from the transaction. Ultimately, the target UE selects the most reliable candidate from the pool based on their contribution history and blockchain-based transaction verification. Conversely, the delivery of content between devices and the edge union is realized by offloading tasks to the edge. To safeguard user privacy, encrypted parameters of computational models are delivered instead of the direct transmission of raw user data, which can effectively prevent data overload and the leakage of sensitive information.

Efficient content caching

Storing frequently accessed content closer to the network edge and caching it in advance can greatly decrease the time it takes for content to be delivered and alleviate traffic jams on the network. Clearly, the key issue in content caching is deciding which content to store at the edge. Hence, precise and rapid estimation of content popularity is crucial for enhancing the performance of content caching. Consider this: a transfer learning-enhanced distributed collaborative filtering approach is proposed here. The reason is that transfer learning can draw on knowledge from other data-rich domains to supply the target domain with substantial and relevant information, effectively tackling the data sparsity issue. To be more specific, unsupervised learning is initially utilized to identify source domains that are similar to the target domain in terms of their attributes. Subsequently, given the large and sparse nature of the content popularity matrix obtained from collaborative filtering in the target domain, transfer learning is adopted to glean insights from source domains to estimate content popularity in the target domain. To boost estimation efficiency and prevent network congestion during knowledge transfer, the popularity matrix is locally computed rather than raw data delivery. Ultimately, proactive content caching is utilized to enhance energy efficiency and minimize the latency of content delivery.

Collaborative computation offloading

The primary role of computation offloading is offering computational services to resource-limited UE to complete their tasks. Inspired by the fact that nearly all UEs possess some computational capabilities, a new collaborative computation offloading scheme is put forward here, leveraging the combined computational power of UEs and the edge union. To be more specific, the computing task is divided into two segments based on the early-exit mechanism [43], with only the most computation-intensive segment being offloaded to the edge union at a minimal content delivery cost. The remaining part is processed on the local UE to ensure energy efficiency and reduced latency. To illustrate collaborative computation in more detail, a deep neural network (DNN) is selected. A segment of the DNN, split at the exit points [43], is initially executed on the local UE to produce certain model-specific parameters. These encrypted parameters are then sent to the edge union for additional computation over wireless channels. Ultimately, the edge union completes the task and sends the computation outcomes back to the UE. This collaborative scheme is well-equipped to support emerging delay-sensitive applications such as haptic-dominant services and equally important multi-modal services. It is essential to note that the decision on where to partition the model hinges on varying needs for accuracy and computation time.

Adaptive optimization control

Compared with traditional 3C frameworks, a novel term "control" is induced into them for joint optimization about communication, caching, and computation. In other words, an adaptive control mechanism is performed by jointly considering secure communication, efficient caching, and collaborative computation. Here, the end-to-end delay and transmission bandwidth consumption should be minimized to support multi-modal streaming with low latency and large volume bandwidth provision demands. At the same time, energy efficiency and high reliability should also be considered.

To be more specific, one UE is assumed to have U tasks to be processed, that is, $D = \{D_1, D_2, \ldots, D_u\}$. Each task contains audio, visual, and haptic modalities, denoted as $D_u = \{D_u^{(a)}, D_u^{(v)}, D_u^{(h)}\}$. Caching decision variable and offloading to the edge union decision variable are denoted as $\alpha_u^{(c)} \in \{0, 1\}$ and $\beta_u^{(c)} \in [0, 1]$, respectively. Based on whether the content is cached, the end-to-end delay $T_u^{(c)}$ and the consumed bandwidth $B_u^{(c)}$ of $D_u^{(c)}$ are two items of the optimization objective function, that is

$$\min_{\alpha_u^{(c)},\beta_u^{(c)},l_u^{(c)},o_u^{(c)}} \sum_u \sum_c T_u^{(c)}\left(\alpha_u^{(c)}, \beta_u^{(c)}, l_u^{(c)}, o_u^{(c)}\right) + \eta_u^{(c)} B_u^{(c)}\left(\beta_u^{(c)}\right), \tag{4.107}$$

where $l_u^{(c)} \in [0, 1]$ is the percentage of the allocated computing resources from edge to $D_u^{(c)}$. $o_u^{(c)} \in [0, 1]$ is the percentage of the allocated computing resources from local UE to $D_u^{(c)}$. $\eta_u^{(c)} > 0$ is the parameter for unifying the range between the first and second items. Moreover, the constraints impacting optimization control should also be considered: $T_u^{(c)}$ cannot exceed $T_{\max}^{(c)}$; $B_u^{(c)}$ should be larger than $B_{\min}^{(c)}$; The tolerable perceived synchronization latency T_{syn} should be kept at the receiver side.

4.7.2 Reinforcement Learning-Based Solution

To address the non-convex optimization issue previously discussed, a solution based on reinforcement learning (RL) [46] for the edge union is designed, shown in Fig. 4.22. RL is a way for an agent to learn through "trial and error". The agent interacts with the environment to obtain rewards that guide its behavior, with the goal of maximizing the reward. Through this method, the agent acquires knowledge in an action-evaluation environment and improves its action plan to adapt to the environment. In other words, the reasons for choosing this method are that it is suitable for dynamic fluctuations of crucial parameters, such as the popularity of content, the connectivity between UEs and the edge union, etc. It can be adaptive to dynamic network conditions, diverse kinds of multi-modal services, and heterogeneous streams. Importantly, the state, action, and reward components in the RL framework are defined in the following.

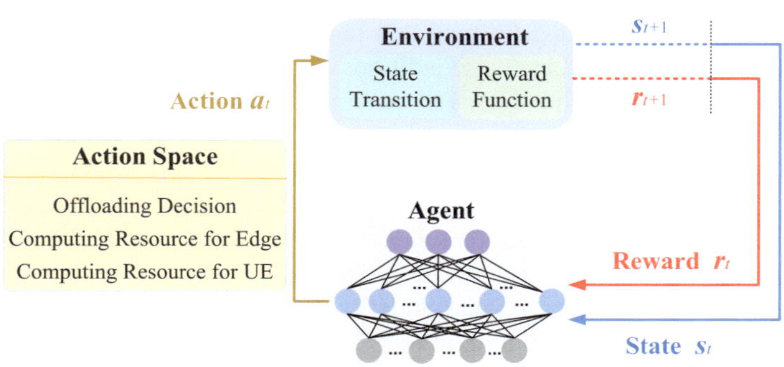

Fig. 4.22 Mechanism of the RL-based solution

State: The state space is defined as $s = \{M_u^{(c)}, T_u^{(c)}, B_u^{(c)}\}$, where each element corresponds to the technical requirements, end-to-end latency, and bandwidth usage of the specific multi-modal service, respectively.

Action: Within the RL framework, the agent will determine the allocation of caching and computing resources for the task at hand, and the proportion of the task that should be offloaded to the edge union. Specifically, the action space can be delineated as: $a = \{\alpha_u^{(c)}, \beta_u^{(c)}, l_u^{(c)}, o_u^{(c)}\}$, encompassing the caching and offloading decision variable, the fraction of computing resources assigned to the task from the edge and local UE.

Reward: The reward function is derived from the objective function presented in Eq. (4.107). To achieve the goal of maximizing cumulative rewards, a negative sign to the objective function is introduced, thereby obtaining the ultimate reward.

To appraise the performance of the proposed framework, experiments about audio-visual-dominant and haptic-dominant multi-modal services are executed. The performance utility is selected as metric and the results are respectively shown in Fig. 4.23a and b. To substantiate the significance of control mechanisms and the influence of each 4C element on the performance, three competing frameworks are taken as benchmarks: caching, computation, and 3C. The quantitative outcomes indicate that the designed 4C framework, integrated with RL, outperforms others under these two patterns of multi-modal services. Moreover, as the 4C elements incrementally increase, the performance utility progressively enhances from caching to 4C. Most crucially, this also suggests that the designed framework is capable of adapting to dynamic network conditions, diverse service requirements, and heterogeneous streams.

4.8 Summary

In this chapter, we have described cross-modal transmission techniques. As analyzed, various service scenarios (i.e., distinguished contents and requests for different categories of multi-modal services), heterogeneous streams (i.e., different transmission requirements for audio-visual and haptic streams), and dynamic network environments (i.e., varied channel conditions and limited transmission resources) are three main issues to be handled. The existing cross-modal transmission schemes belong to either a preemptive-resume pattern or a non-preemptive-resume scheduling pattern. For the former, haptic streams have the highest priorities during transmission. We introduce three representative solutions: the intra-modal correlation-based haptic streaming, the inter-modal correlation-based audio-visual streaming, and the content-aware stream scheduling schemes. The core idea is to leverage the correlations within one modality or among modalities to guarantee low latency and high reliability of haptic streaming, or to relieve competition between haptic and audio-visual streams. Experimental results evaluate the effectiveness of these schemes. For the latter, the requirements of different multi-modal services and characteristics of different modalities are comprehensively considered. We introduce four representative solutions: the modal-aware stream scheduling, the prediction-aware

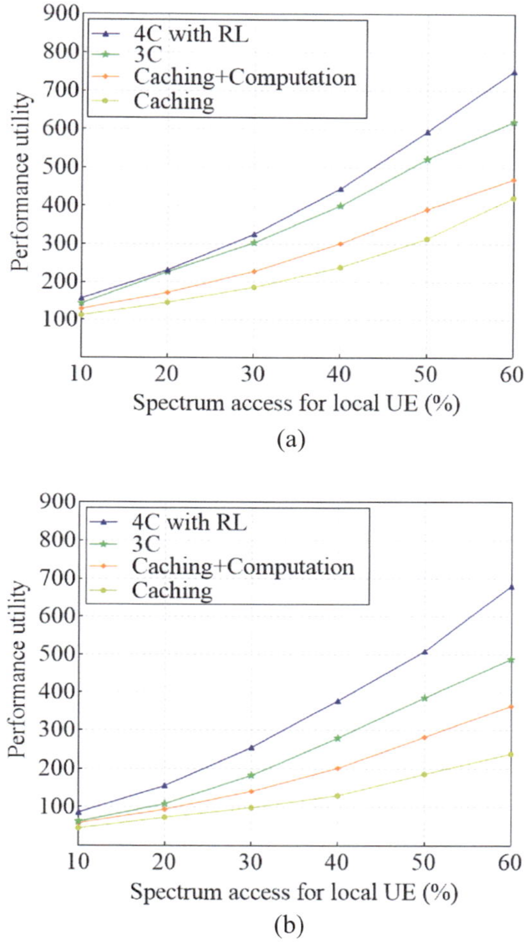

Fig. 4.23 The performance of the proposed framework: **a** Performance utility on the audio-visual dominant service; **b** Performance utility on the haptic-dominant service

stream scheduling, the dynamic transmission mode selection scheme, and the edge intelligence-empowered transmission decision scheme. The core idea is to realize efficient cross-modal transmission by designing targeted scheduling schemes and scheduling switch mechanisms, or predicting the upcoming packets and delivering them to the receiver in advance, or jointly selecting transmission mode and allocating resources, or applying edge intelligence-based 4C optimization. From experimental results, these approaches can fulfill the transmission requirements of different modality streams according to the specific category of multimodal services. They can also effectively adapt to the variations of the transmission environments.

References

1. Yang L, Wu D, Zhou L (2021) Heterogeneous stream scheduling for cross-modal transmission. IEEE Trans Commun 69(9):6037–6049
2. Hou Z, She C, Li Y, Zhuo L, Vucetic B (2020) Prediction and communication co-design for ultra-reliable and low-latency communications. IEEE Trans Wireless Commun 19(2):1196–1209
3. Chang CS, Thomas JA (1995) Effective bandwidth in high-speed digital networks. IEEE J Sel Areas Commun 13(6):1091–1110
4. Liang L, Li GY, Xu W (2017) Resource allocation for D2D-enabled vehicular communication. IEEE Trans Commun 65(7):3186–3197
5. Cizmeci B, Xu X, Chaudhari R, Alt N, Steinbach E (2017) A multiplexing scheme for multimodal teleoperation. ACM Trans Multim Comput 13(2):21
6. Simsek M, Aijaz A, Dohler M, Sachs J, Fettweis G (2016) 5G-enabled tactile internet. IEEE J Sel Areas Commun 34(3):460–473
7. Chen M, Hao Y, Hu L, Huang K, Lau V (2017) Green and mobility-aware caching in 5G networks. IEEE Trans Wireless Commun 16(12):8347–8361
8. Gao Y, Ni S, Wu D, Zhou L (2022) Edge-based cross-modal communications for remote healthcare. IEEE J Sel Areas Commun 40(11):3139–3151
9. Gao Y, Wei X, Zhou L (2020) Personalized QoE improvement for networking video service. IEEE J Sel Areas Commun 38(10):2311–2323
10. Chen M, Qian Y, Hao Y, Li Y, Song J (2018) Data-driven computing and caching in 5G networks: architecture and delay analysis. IEEE Wireless Commun 25(1):70–75
11. Anand A, Veciana GD, Shakkottai S (2020) Joint scheduling of URLLC and eMBB traffic in 5G wireless networks. IEEE/ACM Trans Netw 28(2):477–490
12. Steinbach E, Hirche S, Ernst M, Brandi F, Chaudhari R, Kammerl J, Vittorias I (2012) Haptic communications. Proc IEEE 100(4):937–956
13. Yuan Z, Wei X, Zhou L, Zhuang W (2024) Content-aware cross-modal stream transmission. IEEE Wirel Commun Lett 13(9):2507–2511
14. Polyanskiy Y, Poor HV, Verdu S (2010) Channel coding rate in the finite blocklength regime. IEEE Trans Inform Theory 56(5):2307–2359
15. Stolyar AL (2005) On the asymptotic optimality of the gradient scheduling algorithm for multiuser throughput allocation. Oper Res 53(1):12–25
16. Yuan Z, Kang B, Wei X, Zhou L (2022) Exploring the benefits of cross-modal coding. IEEE Trans Circ Syst Vid 32(12):8781–8794
17. Zhou L, Wu D, Wei X, Chen J (2021) Cross-modal stream scheduling for eHealth. IEEE J Sel Areas Comm 39(2):426–437
18. Yan Z, Liu Q, Zhang T, Chen CW (2018) CrowdDBS: a crowdsourced brightness scaling optimization for display energy reduction in mobile video. IEEE Trans Mobile Comput 17(11):2536–2549
19. Bello O, Zeadally S (2016) Intelligent device-to-device communication in the internet of things. IEEE Syst J 10(3):1172–1182
20. Xu C, Wang M, Chen X, Zhong L, Grieco LA (2018) Optimal information centric caching in 5G device-to-device communications. IEEE Trans Mobile Comput 17(9):2114–2126
21. Wu D, Zhou L (2024) Cross-modal stream transmission: architecture, strategy, and technology. IEEE Wireless Commun 31(1):134–140
22. Chen Y, Li P, Li A, Wu D, Zhou L, Qian Y (2024) Toward low-latency cross-modal communication: a flexible prediction scheme. IEEE Trans Mobile Comput 23(2):13110–13324
23. Ren S, He K, Girshick R B, Sun J (2015) Faster R-CNN: towards real-time object detection with region proposal networks. In: Proceedings of the international conference computer vision, Santiago, Chile, Dec 13–17, pp 27–34
24. Wang S, Li C, Li Y, Yuan Y, Wang G (2023) Self-supervised information bottleneck for deep multi-view subspace clustering. IEEE Trans Image Process 32:1555–1567

25. Zhang L, Zhang M, Song R, Zhao Z, Li X (2023) Unsupervised embedding learning with mutual-information graph convolutional networks. IEEE Trans Multimed 25:5916–5926
26. Sheng Z, Xu Y, Xue S, Li D (2022) Graph-based spatial-temporal convolutional network for vehicle trajectory prediction in autonomous driving. IEEE Trans Intell Transp Syst 23(10):17654–17665
27. Wang T, Yang C, Kirchner F, Du P, Sun F, Fang B (2019) Multimodal grasp data set: a novel visual-tactile data set for robotic manipulation. Int J Adv Robot Syst 16(1):172988141882157
28. Hochreiter S, Schmidhuber J (1997) Long short-term memory. Neural Comput 9(8):1735–1780
29. Zheng C, Fan X, Wang C, Qi J (2020) GMAN: A graph multi-attention network for traffic prediction. In: Proceedings of the AAAI conference artificial intelligence, Online, Feb 7–12, pp 1234–1241
30. Vaswani A, Shazerr N, Parmar N, Uszkoreit J, Jones L, Gomez A. N, Kaiser L (2017) Attention is all you need. In: Proceedings of the conference workshop neural information processing systems, Long Beach, USA, Dec 4–9, pp 1–14
31. Suo Y, Chen Y, Gao Y, Wei X (2023) Dynamic transmission mode selection for multi-modal services. IEEE Commun Lett 27(3):911–915
32. Kourtellaris CK, Psomas C, Krikidis I (2020) Finite blocklength analysis of multiple access channels with/without cooperation. IEEE Trans Commun 68(10):6317–6330
33. Tajallifar M, Ebrahimi S, Javan MR, Mokari N, Chiaraviglio L (2022) Energy-efficient task offloading under E2E latency constraints. IEEE Trans Commun 70(3):1711–1725
34. Zhou L, Yang Z, Wang H, Guizani M (2014) Impact of execution time on adaptive wireless video scheduling. IEEE J Sel Area Commun 32(4):760–772
35. Fu W, Landman A, van Paassen MM, Mulder M (2018) Modeling human difference threshold in perceiving mechanical properties from force. IEEE Trans Hum-Mach Syst 48(4):359–368
36. Ge X, Zhou R, Li Q (2020) 5G NFV-based tactile internet for mission critical IoT services. IEEE Internet Things 7(7):6150–6163
37. Li J, Li Y, Wang Y (2021) Fuzzy inference NSGA-III algorithm-based multi-objective optimization for switched reluctance generator. IEEE Trans Energy Conver 36(4):3578–3581
38. Kumar P, Shamrao C M, Mukherjee M (2019) Sum-rate maximization with joint power allocation and mode selection in D2D-enabled 5G cellular networks. In: Proceedings of the IEEE international conference on communication, Shanghai, China, May 20–24, pp 1–6
39. Cao Y, Jiang T, He M, Zhang J (2016) Device-to-device communications for energy management: a smart grid case. IEEE J Sel Area Commun 34(1):190–201
40. Asuhaimi FA, Bu S, Nadas JPB, Imran MA (2019) Delay-aware energy-efficient joint power control and mode selection in device-to-device communications for freedom systems in smart grids. IEEE Access 7:87369–87381
41. Li L, Shi D, Hou R, Chen R, Lin B, Pan M (2020) Energy-efficient proactive caching for adaptive video streaming via data-driven optimization. IEEE Internet Things 7(6):5549–5561
42. Chen M, Hao Y, Hu L, Hossain MS, Ghoneim A (2018) Edge-CoCaCo: toward joint optimization of computation, caching, and communication on edge cloud. IEEE Wirel Commun 25(3):21–27
43. Zhou Z, Chen X, Li E, Zeng L, Luo K, Zhang J (2019) Edge intelligence: paving the last mile of artificial intelligence with edge computing. Proc IEEE 107(8):1738–1762
44. Gao Y, Wei X, Kang B, Chen J (2021) Edge intelligence empowered cross-modal streaming transmission. IEEE Netw 35(2):236–243
45. Yang Q, Liu Y, Chen T, Tong Y (2019) Federated machine learning: concept and application. ACM Trans Intell Syst Technol 10(2):1–19
46. Sutton RS, Barto AG (1998) Reinforcement learning: an introduction. MIT Press, Cambridge, MA, USA

Chapter 5
Cross-Modal Signal Recovery

Abstract In order to handle issues about the lost, damaged, or delayed arrival of different modality signals after transmission, the corresponding signal recovery schemes should be designed. Firstly, for haptic-dominant multi-modal services, an audio-visual-aided haptic signal recovery scheme is proposed, while for audio-visual-dominant multi-modal services, a haptic-aided visual signal recovery scheme is presented. The core ideas are to utilize the one modality signal (audio-visual or haptic) to recover the other desired modality signal (haptic or visual) by leveraging their potential correlations. Then, for equally important multi-modal services, a visual-haptic mutual signal recovery scheme is provided, which can be seen as the general case for the two schemes mentioned above. Subsequently, besides these three representative cross-modal signal recovery solutions, information retrieval is another method with fast response and high-reliability characteristics. Under this circumstance, an information retrieval-based signal recovery scheme is designed. Finally, in order to enhance the display quality of the receiver and guarantee human experience, a cross-modal super-resolution reconstruction strategy is proposed, which achieves high-resolution visual signals from the received low-resolution visual signals and the related haptic signals.

5.1 Audio-Visual-Aided Haptic Signal Recovery

In the haptic-dominant multi-modal services, touch sensory information plays a more important role than those of auditory and visual sensories. But in these services, there may face two main technical challenges. On one hand, compared with continuously collected and delivered audio and visual signals, haptic signals have characteristics about bursty and unpredictable emergences. Although cross-modal coding and cross-modal transmission techniques have been employed, haptic stream is still sensitive to interference and noise during sending and delivery. This is extremely fatal for services such as remote industrial robotic grasping, tele-surgery or diagnosis, etc. On the other hand, for services that have no haptic acquisition devices at the slave domain but the touch sensory is needed by humans, i.e. online immersive shopping or gaming, holographic museum guide, there should generate "virtual" touch sensory

X. Wei et al., *Cross-Modal Communication Technology*, Wireless Networks,
https://doi.org/10.1007/978-3-032-00957-9_5

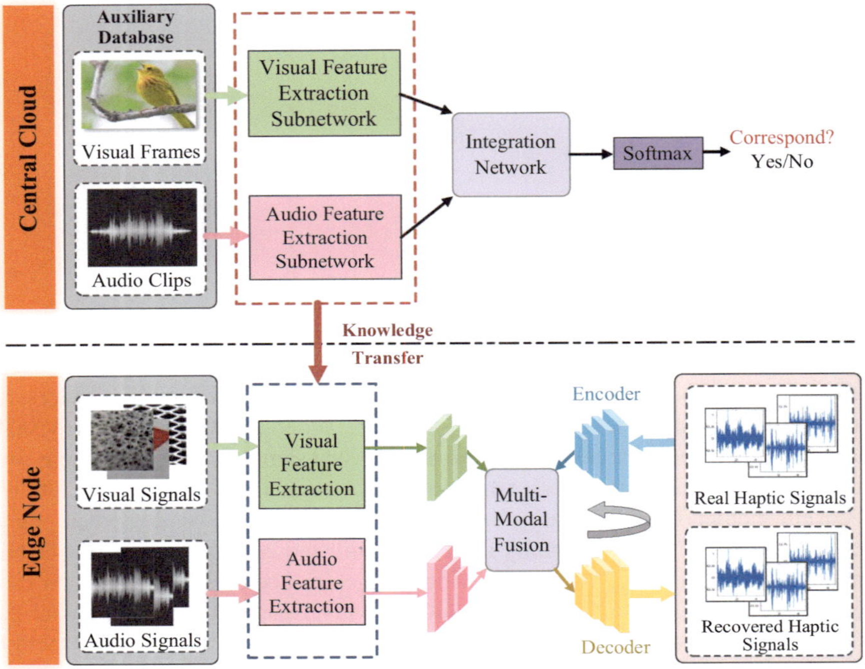

Fig. 5.1 Framework of the AVHR

or haptic signals at the master domain. To tackle these challenges, an audio-visual-aided haptic signal recovery (AVHR) scheme is proposed [1]. It emphasizes the concurrent presence of audio-visual modalities within haptic-dominant multi-modal services and attempts to utilize the potential correlation among audio-visual and haptic modalities to achieve haptic signal recovery. Framework of the AVHR is shown in Fig. 5.1, constructed on the cloud-edge collaboration-based framework. It can be further divided into three main modules: feature extraction by cloud-edge transfer, shared semantic learning by multi-modal fusion, and haptic signal generation by semantic constraints. In general, the central cloud leverages its auxiliary audio-visual database to acquire knowledge and subsequently conveys this knowledge to the edge node at the master domain. The edge node integrates the locally received audio and visual signals with the knowledge transferred from the central cloud, thoroughly exploring the potential connections and consistency across modalities to ultimately recover the desired haptic signals.

5.1.1 Feature Extraction by Cloud-Edge Transfer

The first module of the AVHR involves extracting features from the audio-visual signals. However, this process faces certain challenges. Firstly, the amount of available audio and visual signals for model training at the edge node is limited. Secondly, it is impractical to manually annotate all the audio-visual signals in the training dataset with semantic labels. To address the first issue, in addition to the audio-visual dataset at the edge node, an auxiliary database with a vast amount of unlabeled audio-visual data stored in the central cloud can be utilized to distill and provide semantic knowledge. To tackle the second problem, self-supervised learning can be employed to get audio-visual representations.

Specifically, let the auxiliary audio-visual database denote by $S = \{s_j\}_{j=1}^{N}$, where $s_j = (v_j^s, a_j^s)$. The unlabeled visual frames v_j^s and audio clips a_j^s are respectively input into the visual and audio subnetworks, obtaining visual and audio features. Features from these two modalities are then concatenated and into a multi-layer neural network. Finally, a self-supervised learning task can be employed to the output of the neural network. The task is to judge whether a_j^s and v_j^s belongs to the identical audio-visual source or not [2]. Consider this, the loss of this task can be designed as

$$\mathcal{L}_{Src} = -\frac{1}{N}\sum_{j=1}^{N} f_{sig}\left(g_j^s, p\left(g_j^s\right); \theta_g^s\right), \tag{5.1}$$

where $g_j^s = [G_v(v_j^s; \theta_v^s), G_a(a_j^s; \theta_a^s)]$ is the feature after audio-visual modality concatenation, $G_v(\cdot)$, $G_a(\cdot)$ are the feature mapping of visual subnetwork and audio subnetwork with θ_v^s and θ_a^s, respectively. $p(\cdot)$ represents the label indicator. If a_j^s and v_j^s belong to the identical source, $p(g_j^s) = 1$; otherwise, $p(g_j^s) = 0$. Additionally, θ_g^s is the parameters of the multi-layer neural network, and $f_{\text{sig}}(\cdot)$ represents the cross entropy loss of sigmoid.

By minimizing $\mathcal{L}_{Src}$, $G_v(\cdot)$, $G_a(\cdot)$, θ_v^s, θ_a^s can be estimated, which are taken as "knowledge" and transferred to the feature extraction network at the edge node. In other words, the edge node can take advantage of this "knowledge" for the start point of its own feature extraction network.

5.1.2 Shared Semantic Learning by Multi-modal Fusion

Although the high degree of heterogeneity among the received audio-visual signals and haptic signals to be recovered, they share the same semantics or have correlation. As a result, it is essential to develop an effective method to explore these semantics, which is beneficial to the final haptic signal generation.

Let $D = \{d_i\}_{i=1}^{M}$, $d_i = (v_i, a_i, h_i)$ denote the M visual-audio-haptic signal pairs for model training. For the i-th instance d_i, $v_i \in R^v$, $a_i \in R^a$, and $h_i \in R^h$. More-

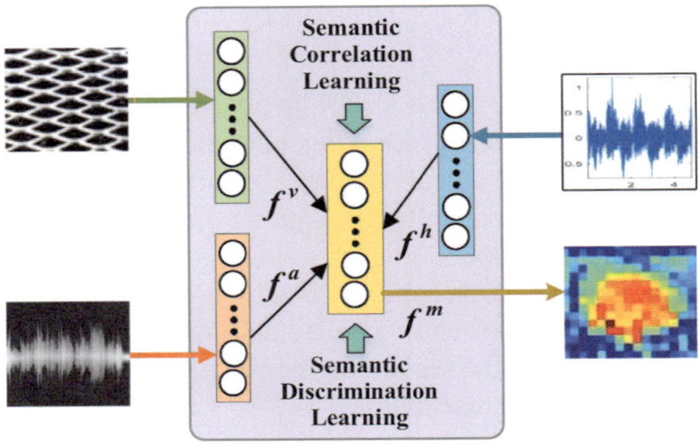

Fig. 5.2 The shared semantic learning paradigm by multi-modal fusion

over, the head of each received packet can provide the related category label. It is taken as one-hot coding $y_i \in R^K$, where K is the number of categories. Specifically, $g^v = G_v\left(v; \theta_v^s\right)$ and $g^a = G_a\left(a; \theta_a^s\right)$ are firstly derived from the transferred visual and audio feature extraction networks. Then, g^v and g^a are further input to their respective multi-layer neural networks to obtain the features with the same dimension, $f^v = F_v\left(g^v; \theta_v\right)$ and $f^a = F_a\left(g^a; \theta_a\right)$. At the same time, haptic feature with the same dimension can also be derived from the corresponding haptic signal, $f^h = E_h\left(h_i; \theta_{he}\right)$. $E_h(\cdot)$ is the encoder part of an autoencoder model, explained in Sect. 5.1.3. θ_v, θ_a and θ_{he} are the related parameters.

After obtaining f^a, f^v, f^h, the most key step is to combine collaborative and joint paradigms for exploring the shared semantics, depicted in Fig. 5.2.

Semantic correlation learning

Here, multi-modal collaborative paradigm [3, 4] is firstly considered, which achieves collaboration of audio and visual modalities by imposing the constraints of the haptic modality. In other words, its object is to maximize semantic similarity among f^a, f^v, and f^h. Furthermore, joint paradigm is accumulated, which fusing audio and visual modalities together,

$$f^m = F_m\left(f^a, f^v; \theta_m\right), \tag{5.2}$$

where f^m is the fusion output in the shared semantic subspace, $F_m(\cdot)$ is the mapping function with parameter θ_m.

To derive the shared semantics, it is necessary for f^a, f^v, and f^m to be similar to f^h. This can be realized by designing the contrastive loss, which enforces the correlation among modalities. Specifically, it tries to minimize the distance between matching feature pairs while ensuring that the distance between mismatching feature pairs exceeds a certain threshold δ. The loss functions are formulated as

$$\mathcal{L}_{Corr}^{av} = \sum_{p \neq q}^{M,M} \max\left(0, l_2^2\left(f_p^a, f_p^h\right) + l_2^2\left(f_p^v, f_p^h\right) + \delta - l_2^2\left(f_p^a, f_q^h\right) - l_2^2\left(f_p^v, f_q^h\right)\right), \tag{5.3}$$

$$\mathcal{L}_{Corr}^{m} = \sum_{p \neq q}^{M,M} \max\left(0, l_2^2\left(f_p^m, f_p^h\right) + \delta - l_2^2\left(f_p^m, f_q^h\right)\right), \tag{5.4}$$

where $\mathcal{L}_{Corr}^{av}$ and $\mathcal{L}_{Corr}^{m}$ are contrast losses about f^a, $f^v \sim f^h$ and $f^m \sim f^h$ pairs. $l_2 = \|\cdot\|_2$ is l_2-norm. Based on these two losses, the objective function about semantic correlation learning can be constructed as

$$\mathcal{L}_{Corr} = \mathcal{L}_{Corr}^{av} + \mathcal{L}_{Corr}^{m}. \tag{5.5}$$

By minimizing $\mathcal{L}_{Corr}$, the optimal θ_v, θ_a, θ_{he} and θ_m in $F_v(\cdot)$, $F_a(\cdot)$, $E_h(\cdot)$ and $F_m(\cdot)$ can be estimated.

Semantic discrimination learning

In addition to semantic correlation learning, training signals (f^a, f^v, f^h, f^m) from the same category should exhibit greater semantic similarity compared to those from different categories. This can be realized with the help of category supervision information. Specifically, a full connected layer with a softmax function is taken as the common classifier, and the corresponding semantic discrimination loss function is formulated as

$$\mathcal{L}_{Dis} = -\frac{1}{M}\sum_{i=1}^{M} y_i \left[\log p\left(f_i^v; \theta_l\right) + \log p\left(f_i^a; \theta_l\right) + \log p(f_i^h; \theta_l) + \log p\left(f_i^m; \theta_l\right)\right], \tag{5.6}$$

where $p(\cdot)$ denotes the classifier, y_i is the corresponding real label, and θ_l is the parameter of the common classifier. By optimizing $\mathcal{L}_{Dis}$, the semantic differences between features from different categories can be maximized while the differences between features within the same category can be minimized.

In summary, by carefully designing $\mathcal{L}_{Corr}$ and $\mathcal{L}_{Dis}$, the semantics in features after fusion can be explored, which will determine the quality of the recovered haptic signals.

5.1.3 Haptic Signal Generation by Semantic Constraints

Following the exploration of shared semantics, the ultimate module is to recover or generate haptic signals. As discussed in Sect. 5.1.2, an autoencoder is employed, with the encoder $E(\cdot)$ responsible for haptic feature extraction and the decoder $D(\cdot)$ utilized for haptic signal generation. Its training process contains the following two main processes.

The first process is to estimated parameters of the autoencoder by minimizing the difference between the output haptic signals from the decoder and the real haptic signals. The loss function is formulated as

$$\mathcal{L}_{Rec} = \frac{1}{M}\sum_{i=1}^{M}\left\|\widetilde{h}_i - h_i\right\|_2^2 + \alpha\left\|\theta_h\right\|_2^2, \tag{5.7}$$

where h_i is the real haptic signal. $\widetilde{h}_i = D_h\left(E_h\left(h_i;\theta_{he}\right);\theta_{hd}\right)$ is the output haptic signal. $\theta_h = [\theta_{he}, \theta_{hd}]$ denote the parameters of the encoder and the decoder. By optimizing $\mathcal{L}_{Rec}$, θ_{hd} and θ_{he} can be estimated. After θ_{he} is updated, the reasonability of haptic feature f^h from the output of encoder will further improved.

The second process is to generate the desired haptic signal h' by inputting the fused feature f^m into $D_h(\cdot)$. When concerning the design of loss function, it can be further divided into three aspects. To form the structure of the generated haptic signal h', $\left\|h' - h\right\|_2$ should be imposed. Moreover, to guarantee the semantic integrity, h' is re-input into $E_h(\cdot)$, outputting haptic feature $f^{h'}$. Based on this, feature semantic similarity between $f^{h'}$ and f^h should be considered. Furthermore, category discrimination of $f^{h'}$ can also take effect for the generation process. Taking into account the three aspects mentioned above, the loss function about the desired haptic signal generation can be formulated as

$$\mathcal{L}_{Gen} = \frac{1}{M}\sum_{i=1}^{M}\left\{\left\|h_i - h_i'\right\|_2^2 + \beta\left[l_2^2\left(f_i^h, f_i^{h'}\right) + y_i\log\left(f_i^{h'}\right)\right]\right\} + \gamma\left\|\theta_{hd}\right\|_2^2, \tag{5.8}$$

where $h_i' = D_h(f_i^m;\theta_{hd})$. $f_i^{h'} = E_h(h_i';\theta_{he})$ is the semantic feature of h_i' extracted by the encoder. $l_2(f_i^h, f_i^{h'})$ denotes the similarity of features f_i^h and $f_i^{h'}$. $y_i\log(f_i^{h'})$ is the classification loss of $f_i^{h'}$. By optimizing $\mathcal{L}_{Gen}$, the result can be more structured, ensuring semantic integrity and thereby generating high-quality haptic signals.

5.1.4 Optimization Process

After designing the three modules in the AVHR, the optimization process should be performed. In summary, the related loss functions are listed in the following and the specific process is shown in Algorithm 5.1. The parameters in the loss functions are estimated by the stochastic gradient descent (SGD) algorithm.

$$
\begin{aligned}
&\text{cloud}: \mathcal{L}_{Src}\left(\theta_v^s, \theta_a^s, \theta_g^s\right) \\
&\text{edge}: \mathcal{L}_{Corr}\left(\theta_v, \theta_a, \theta_{he}, \theta_m\right) + \mathcal{L}_{Dis}\left(\theta_v, \theta_a, \theta_{he}, \theta_m, \theta_l\right) \\
&\qquad + \mathcal{L}_{Rec}\left(\theta_{he}, \theta_{hd}\right) + \mathcal{L}_{Gen}\left(\theta_{hd}\right).
\end{aligned} \tag{5.9}
$$

Algorithm 5.1 Optimization process of the AVHR [1]

Require: Auxiliary database S on the cloud, training dataset D at the edge;
Initial parameters $\theta_v^s, \theta_a^s, \theta_g^s$ at the central cloud and $\theta_v, \theta_a, \theta_m, \theta_l, \theta_{he}, \theta_{hd}$ at the edge node;
Batch-size, learning rate μ_1, μ_2, iterations n_1, n_2

Stage 1: feature extraction by cloud-edge transfer
for iteration $= 1, \ldots, n_1$ **do**
  Update θ_v^s, θ_a^s, and θ_g^s by SGD for $\mathcal{L}_{Src}$
  $\theta_v^s \leftarrow \theta_v^s - \mu_1 \nabla_{\theta_v^s} \mathcal{L}_{Src}$
  $\theta_a^s \leftarrow \theta_a^s - \mu_1 \nabla_{\theta_a^s} \mathcal{L}_{Src}$
  $\theta_g^s \leftarrow \theta_g^s - \mu_1 \nabla_{\theta_g^s} \mathcal{L}_{Src}$
end for
return $G_v\left(\theta_v^s\right), G_a\left(\theta_a^s\right)$, transfer them from cloud to edge
Stage 2: shared semantic learning & haptic signal generation
for iteration $= 1, \ldots, n_2$ **do**
  Update $\theta_v, \theta_a, \theta_m, \theta_{he}, \theta_l$ by SGD for $\mathcal{L}_{Corr}$ and $\mathcal{L}_{Dis}$
  $\theta_v \leftarrow \theta_v - \mu_2 \nabla_{\theta_v}\left(\mathcal{L}_{Corr} + \mathcal{L}_{Dis}\right)$
  $\theta_a \leftarrow \theta_a - \mu_2 \nabla_{\theta_a}\left(\mathcal{L}_{Corr} + \mathcal{L}_{Dis}\right)$
  $\theta_m \leftarrow \theta_m - \mu_2 \nabla_{\theta_m}\left(\mathcal{L}_{Corr} + \mathcal{L}_{Dis}\right)$
  $\theta_{he} \leftarrow \theta_{he} - \mu_2 \nabla_{\theta_{he}}\left(\mathcal{L}_{Corr} + \mathcal{L}_{Dis}\right)$
  $\theta_l \leftarrow \theta_l - \mu_2 \nabla_{\theta_l} \mathcal{L}_{Dis}$
  Update θ_{he} and θ_{hd} by SGD for $\mathcal{L}_{Rec}$ and $\mathcal{L}_{Gen}$
  $\theta_{he} \leftarrow \theta_{he} - \mu_2 \nabla_{\theta_{he}} \mathcal{L}_{Rec}$
  $\theta_{hd} \leftarrow \theta_{hd} - \mu_2 \nabla_{\theta_{hd}}\left(\mathcal{L}_{Rec} + \mathcal{L}_{Gen}\right)$
end for
Output: $F_v\left(\theta_v\right), F_a\left(\theta_a\right), F_m\left(\theta_m\right), E_h\left(\theta_{he}\right), D_h\left(\theta_{hd}\right)$

It should be highlighted that the latter two components of the AVHR, deployed at the edge node, constitute an end-to-end network framework. These components are trained jointly and iteratively until convergence is achieved. Once the parameter optimization or model training is complete, the AVHR is automatically triggered and executed at the edge node under specific conditions. These conditions include the detection of missing or impaired haptic packets or the need to generate "virtual" haptic signals. In such scenarios, the received audio and visual signals are decoded and fed into their respective feature extraction networks. Subsequently, the semantic features extracted from these two modalities are fused and input into the trained decoder of the haptic-related autoencoder to generate the desired haptic signals.

5.1.5 Experimental Result and Analysis

In this section, the AVHR approach is employed on a standard dataset. Its performance can be evaluated by comparing with the other competing approaches.

Experimental configuration

Dataset: It contains both a large-scale video database [5] located at the central cloud, and an audio-visual-haptic dataset LMT [6] deployed at the edge node. For the former, it has a high degree of audio-visual correspondence. For each video, 256 × 256 visual frames and 1 second audio clips are sampled. A total of 80 million samples are gathered for exploring "knowledge". For the latter, it is about 108 surface materials with 9 categories (mesh, stone, metal, wood, rubber, fiber, foam, foil and paper, textiles and fabrics). The ratio of training set and testing set volume is about 4:1.

Competing schemes: Here, four competing schemes are selected, including JEC-GAN [7], DVHL [8], E-GAN [9], and SR. Specifically, the joint-encoding-classification GAN (JEC-GAN) utilizes distinct coding networks to encode instances from each modality into a shared intrinsic space. Haptic signals are generated from the input visual information and the obtained semantics in the shared intrinsic space. The deep visual-haptic learning (DVHL) builds upon the traditional autoencoder by incorporating latent variables about visual-haptic correlation. The ensembled GAN (E-GAN) uses the texture attribute of an image as input to derive its associated category information and performs cross-modal conversion to haptic spectra. The simple regression (SR) denotes the most basic generation network, which takes visual information as input and haptic signals as the target output, without imposing any additional constraints.

Implementation details: The implementation details of all the modules and networks in the AVHR are listed in the following.

Visual subnetwork (at the cloud): For the input 256 × 256 visual signals. The VGG network design style is adopted, which includes a 2 × 2 convolution filter and a 2 × 2 max-pooling layer with a stride of 2 and no padding. The network is divided into four blocks, each containing two convolution layers and one pooling layer. The number of filters doubles in each consecutive block. Finally, max-pooling is applied across all spatial locations to produce a single 512-dimensional visual feature vector.

Audio subnetwork (at the cloud): An 1 s audio clip is initially transformed into a linear spectrogram. The structure of the audio subnetwork is similar to that of the visual subnetwork, except that its input is one-dimensional intensity data, while the output is a 512-dimensional feature vector.

Integration network (at the cloud): The 512-dimensional visual and audio feature vectors are concatenated to form an 1024-dimensional vector. This vector is then passed through an integration network consisting of two fully connected layers (128-2) to generate a binary classification output.

Audio-visual feature extraction & multi-modal fusion network (at the edge): It retains the structure and parameters of the visual and audio subnetworks at the cloud. A three-layer fully connected neural network (256-128-32) is subsequently constructed. Here, the visual and audio features are represented as the two 32-dimensional vectors. These vectors are fed into a two-layer feed-forward neural network (64-32) for semantic fusion. Additionally, to perform the semantic discrimination learning described in Sect. 5.1.2, a fully connected layer with K nodes and a softmax function is used as a common classifier to output the specific category labels of audio, visual, haptic, and fused features.

Haptic-related autoencoder (at the edge): This is constructed based on a stacked autoencoder with symmetrical encoder and decoder structures. The encoder consists of a four-layer feed-forward neural network that projects haptic signals into semantic features and labels (Z-256-128-32), where Z represents the dimension of the input haptic signal. The decoder has the same structure but in reverse order.

Evaluation metrics: Here, three specific metrics are adopted to evaluate the performance of the AVHR approach. Firstly, the root mean square error (RMSE) directly measures the recovery error between the generated haptic signals h_i' and the most similar haptic signals h_i in the testing set,

$$RMSE = \sqrt{\frac{1}{N}\sum_{i=1}^{N}\left(h_i - h_i'\right)^2}, \tag{5.10}$$

where N represents the number of samples. Secondly, SIM is the metric measuring the similarity of h_i' and h_i,

$$SIM = \frac{1}{N}\sum_{i=1}^{N}\frac{1}{e^{\mathrm{dist}\left(h_i, h_i'\right)}}, \tag{5.11}$$

where dist $\left(h_i, h_i'\right)$ is the Chebyshev distance. The range of SIM is [0, 1]. When SIM is closer to 1, it means the two signals are more similar, and vice versa. Thirdly, accuracy (ACC) is utilized whether the category of the generated haptic signals is consistent with the real category.

Results and analysis

We initially compare the proposed AVHR with the other competing approaches. The experimental results are presented in Table 5.1. It can be observed that the AVHR outperforms all competing approaches in recovery performance. The reasons are mainly threefold: (1) Knowledge transfer from the cloud enhances the extraction of audio-visual features at the edge. (2) The fusion of audio-visual features and semantic learning deeply explores their potential correlations. (3) Cyclic optimization of the autoencoder improves the quality of the generated haptic signals.

Additionally, we qualitatively compare the recovery effects of the AVHR approach on five material categories. Figure 5.3 shows the images, sounds, real haptic signals, and the recovered or generated haptic signals using the AVHR, JEC-GAN, and DVHL

Table 5.1 Performance comparison of the AVHR and the other competing approaches

Approach	RMSE	SIM	ACC
SR	2.3788	0.7054	0.4101
E-GAN	3.4702	0.7459	0.4833
DVHL	3.4597	0.7445	0.5140
AVHR	**0.8994**	**0.8384**	**0.7580**

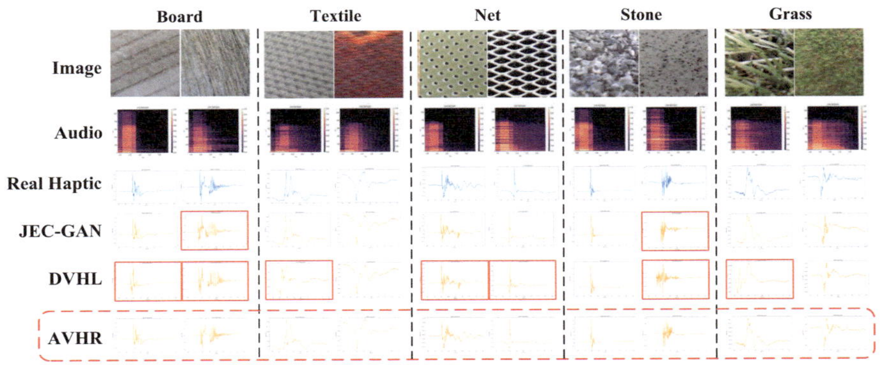

Fig. 5.3 Recovery results of the haptic signals by the AVHR and the other competing schemes (poor results are denoted by red rectangular boxes)

methods. It is evident that several haptic signals generated by the JEC-GAN and the DVHL are not similar to the real ones, particularly for material categories such as net, metal, and board. In contrast, the AVHR generates high-quality haptic signals that closely resemble the real ones, although some differences remain due to inherent variations within each material category. Furthermore, Fig. 5.4 presents the ACC of the three methods for specific material categories. For categories with high-level semantics, such as stone and metal, signal recovery can be challenging and prone to confusion. However, the AVHR still achieves the highest ACC across all categories compared to the JEC-GAN and the DVHL.

Next, the ablation experiment is executed for evaluating the importance of each module in the AVHR. The results are shown in Table 5.2. Specifically, "AVHR w/o $\mathfrak{L}_{Src}$" represents that knowledge extraction and transfer from cloud is not available. From the comparison between "AVHR w/o $\mathfrak{L}_{Src}$" and "AVHR (full)", the transferred feature network from cloud can help enhance the capabilities of feature extraction networks located at the edge. "AVHR w/o $\mathfrak{L}_{Corr}$ &$\mathfrak{L}_{Dis}$" denotes that the audio and visual features are simply accumulated and the shared semantic learning module is not loaded. The comparison result indicates that this module significantly impacts all metrics by narrowing the gap and capturing the shared semantics across modalities. "AVHR w/o $\mathfrak{L}_{Rec}$" denotes that the extraction of haptic features is no longer carried out by the autoencoder, but is instead replaced by features extracted from a specific

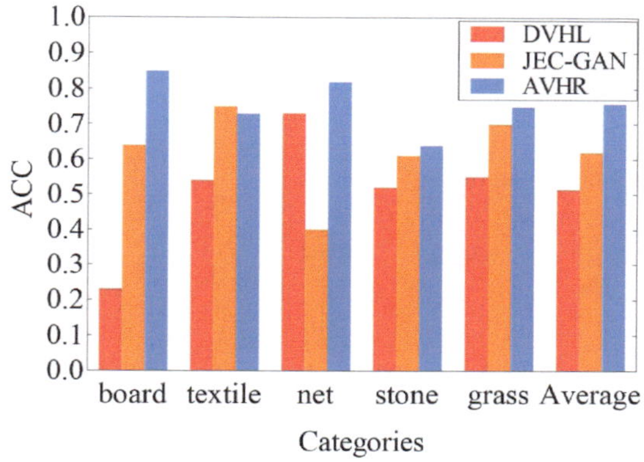

Fig. 5.4 ACC of the AVHR, DVHL, and JEC-GAN on the specific categories

Table 5.2 Result about impacts of each module in the AVHR

Approach	RMSE	SIM	ACC
AVHR w/o $\mathcal{L}_{Src}$	3.2488	0.7207	0.7333
AVHR w/o $\mathcal{L}_{Corr}$ & L_{Dis}	2.1406	0.7478	0.7199
AVHR w/o $\mathcal{L}_{Rec}$	3.3798	0.6885	0.5833
AVHR w/o $\mathcal{L}_{Gen}$	1.8794	0.7748	0.7266
AVHR (fully)	**0.8994**	**0.8384**	**0.7580**

layer of the haptic classification network. After compared with "AVHR (full)", the structure learning is beneficial for exploring haptic semantic representation, thus guaranteeing the quality of the generated haptic signals. "AVHR w/o $\mathcal{L}_{Gen}$" represents that the fused features are directly fed into the decoder, which is trained solely through structure learning. This reflects the necessity of $\mathcal{L}_{Gen}$ in Eq. (5.8). In summary, the above analysis demonstrates that $\mathcal{L}_{Src}$, $\mathcal{L}_{Corr}$ & $\mathcal{L}_{Dis}$, $\mathcal{L}_{Rec}$, and $\mathcal{L}_{Gen}$ all contribute positively to haptic signal recovery.

Finally, to validate the effectiveness of the fusion paradigm or the role of f^m within the AVHR, an additional comparative experiment is conducted. Here, four alternative methods are employed to replace the collaborative-joint fusion paradigm detailed in Sect. 5.1.2. The experimental result is presented in Table 5.3. Specifically, "Only visual modality" approach relies solely on visual signals to recover haptic signals. It leverages cloud-based feature extraction and incorporates haptic features as semantic constraints. The configuration for generating haptic signals is consistent with the AVHR approach. While its recovery performance falls short of the AVHR, it outperforms other competing methods overall. This indicates that visual signals are capable of providing rich semantic information. "Only audio modality" approach involves using solely auditory signals to recover haptic signals, with all the other settings being identical to those in "Only visual modality" scenario. It is observed that this method exhibits the poorest performance among all the tested schemes. "Collabora-

Table 5.3 Result about different fusion paradigms

Approach	RMSE	SIM	ACC
Only visual modality	1.3868	0.8204	0.6833
Only audio modality	3.5786	0.6654	0.6667
Collaborative fusion	2.0632	0.7448	0.7499
Joint fusion	2.2698	0.7712	0.7066
AVHR (collaborative + joint)	**0.8994**	**0.8384**	**0.7580**

tive fusion" involves using haptic features to constrain the extraction of audio-visual features. Subsequently, the extracted audio and visual features are simply concatenated to form the fused feature f^m, which is then used to generate haptic signals. Compared to the "Only audio modality" approach, there is a noticeable improvement in the generation or recovery quality, highlighting the benefits of leveraging complementary information from multiple modalities. However, when compared to the "Only visual modality," collaborative fusion does not demonstrate superior performance in terms of RMSE and SIM. In the "Joint fusion" approach, after extracting 512-dimensional features from both visual and audio signals, the corresponding weights are assigned to these features. These weighted features are then mapped into a common subspace to generate f^m. Compared to "Only visual modality", the haptic signals generated through joint fusion achieve only a marginal improvement in ACC. This suggests that joint fusion has not fully captured the shared semantics across modalities. Based on the results and the preceding analysis, the proposed AVHR approach, which integrates collaborative and joint paradigms for multi-modal fusion, demonstrates superior recovery performance across all metrics.

5.2 Haptic-Aided Visual Signal Recovery

In audio-visual-dominant multi-modal services, the receiving and displaying quality of audio-visual signals is important for satisfying human's immersive experience. Compared with visual streams with continual characteristics, haptic streams have bursty emergence. Due to relatively low transmission priority of visual modality and sudden emergence of the haptic streams, visual streams usually arrive much too late at the receiver, leading to the critical modality synchronization issue. Moreover, due to large-scale size and continual characteristic, visual streaming over wireless channels inevitably experiences packet loss or be subject to interference. Therefore, potential correlation among modalities should be leveraged at the receiver to deal with the challenges about the lost, damaged, or delayed arrival of visual signals. Based on this, a haptic-aided visual signal recovery approach is presented [10]. Its name is the fine-grained haptic to image synthesis (FHIS) and its framework is shown in Fig. 5.5.

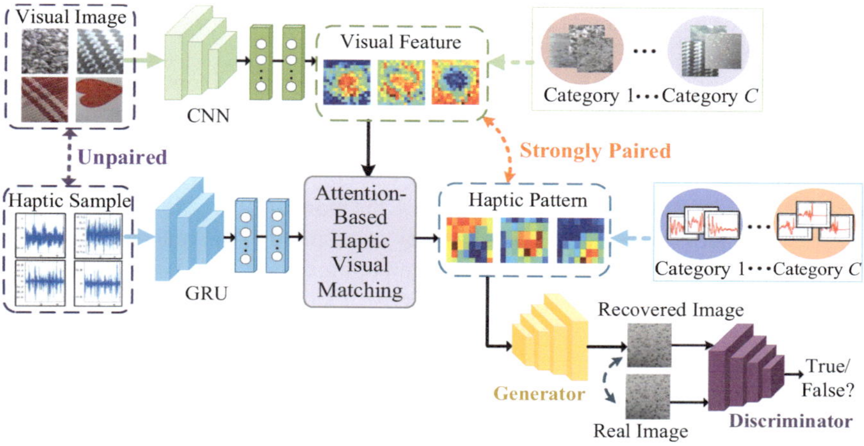

Fig. 5.5 Framework of the FHIS

It mainly contains three modules: low-level discriminative visual feature extraction, middle-level fine-grained haptic-visual semantic association, and high-level global-local image synthesis. In particular, during the training process, the edge node at the receiver selects some previous received haptic and visual signals along with their respective category labels in the packet header for training the FHIS. Initially, the visual images used for training are fed directly into a meticulously designed deep neural network that is supervised by their category labels. This step serves to both reduce computational complexity and extract features that are highly discriminative for different categories. Following this, to uncover the strong correlation between visual and haptic modalities, an attention interaction network is developed. Additionally, the training process incorporates intra-modal discrimination for haptic signal with inter-modal consistency between fine-grained matching haptic-visual pairs, aiming to capture the intrinsic semantics across modalities. Lastly, the visual image synthesis is implemented using generative adversarial networks (GAN), leveraging both "local" pixel matching in images and "global" fine-grained semantics among modalities. Once training is complete, and if visual signal recovery is required at the receiver, the corresponding haptic signal is fed into the trained FHIS to generate the desired visual images.

5.2.1 Low-Level Discriminative Visual Feature Extraction

Here, the received haptic and visual signals for model training are denoted as $H = \{h_1, \ldots, h_j, \ldots, h_m\}$ and $V = \{v_1, \ldots, v_i, \ldots, v_n\}$. h_j represents the j-th haptic sample with dimension d_h, while v_i represents the i-th visual image with dimension d_v. It is noted that due to heterogeneous characteristics, $d_h \neq d_v$. Moreover, each visual image or haptic sample has a category label represented as $y \in R^C$ (C is the total categories).

For extracting valuable visual features from V, a pre-trained convolutional neural network (CNN) with a category label discriminator composed of a fully connected layer and a softmax layer is constructed. It is used to distinguish the authenticity of the predicted category $v^{(c)}$ and the true category label y. The more category attributes the extracted visual features possess, the more difficult it is for the category label discriminator to determine whether they correspond to the true category label. Conversely, as the discriminator is trained and optimized, it becomes more difficult to discern authenticity. Therefore, the feature extraction network needs to achieve higher precision. Through this adversarial training process, the richness of semantic information in the visual feature vectors is enhanced, and a fully connected layer capable of accurately predicting categories is obtained. Consider this, an adversarial loss function is designed,

$$\min_{\theta_V} \max_{\alpha} L_{cat}^{V}(D_1, v^{(c)}) = \mathbb{E}_{v^{(c)}}\big[\log(1 - D_1(v^{(c)}))\big] + \mathbb{E}_y\big[\log D_1(y)\big], \tag{5.12}$$

where $\boldsymbol{\alpha}$ and $\boldsymbol{\theta}_V$ respectively denote parameters of the full-connection layers and the discriminator D_1. After imposing the pre-trained CNN as and the parameter-tuned full connection layers under the supervision of category label, the extracted low-level visual features, denoted as $v^{(f)}$, exhibit relatively abundant intra-modal discriminative category characteristics.

5.2.2 *Middle-Level Fine-Grained Semantic Association*

Upon completing the feature extraction process for visual signals, the subsequent process involves identifying the potential correlations between visual and haptic signals. This process can be further delineated into two primary steps: attention-based haptic-visual modality matching and strongly paired semantic association. The former aims to identify the optimal haptic pattern that correspond to each visual image. The latter focuses on establishing robust semantic association features between the two modalities.

Attention-based haptic-visual modality matching

In cross-modal communications, obtaining strongly paired haptic-visual modality pairs for model training is often challenging due to the unavailability of one-to-one sample correspondence. Moreover, even when a haptic sample and a visual image share the same category labels, their semantic meanings may still be inconsistent or inaccurate. For instance, in the recovery of a "metal mesh" surface material, selecting a visual image of a "metal mesh" and a haptic sample of a "plastic mesh" or "fabric mesh" for training may adversely affect modality semantic mining, ultimately leading to deteriorated recovery outcomes. Additionally, in real-world scenarios, the provided categories for visual and haptic data are often coarse, such as simply labeling them as "mesh". Existing methods have primarily focused on semantic associations between visual and haptic modalities at the coarse category level (e.g. "mesh", "metal", etc.) [11, 12]. To achieve more precise matching of haptic samples with visual images at a finer granularity, the attention mechanism [13] is considered. That is to say, the objective is to identify the most appropriate haptic pattern that corresponds to each visual image.

Specifically, the haptic samples are initially processed through a pre-trained gate recurrent unit (GRU) network followed by several fully connected layers (with $\boldsymbol{\theta}_H$ to be estimated) to extract features. Subsequently, the feature-extracted representations of the candidate haptic samples $\{h_1^{(f)}, \ldots, h_j^{(f)}, \ldots, h_m^{(f)}\}$ and the visual query vector $v_i^{(f)}$ obtained by the module in Sect. 5.2.1 are taken as inputs to the attention network (shown in Fig. 5.6). Following this, the hidden layer representations of the haptic feature vectors, $h_i^{(r)}$, are derived through a single-layer perception with a hyperbolic tangent activation function,

$$h_j^{(r)} = Tanh\left(w_j h_j^{(f)} + b_j\right), \tag{5.13}$$

where w_j and b_j are parameters of the attention network. Based on this, the similarity of $v_i^{(f)}$ and $h_j^{(r)}$ is depicted by

$$\begin{aligned} Sim_{i,j} &= I_{i,j} \frac{h_j^{(r)} \cdot v_i^{(f)} - \left\|h_j^{(r)}\right\| \cdot \left\|v_i^{(f)}\right\|}{\sqrt{\left\|\left(h_j^{(r)}\right)^2\right\| - \left\|h_j^{(r)}\right\|^2} \sqrt{\left\|\left(v_i^{(f)}\right)^2\right\| - \left\|v_i^{(f)}\right\|^2}}, \\ I_{i,j} &= \begin{cases} 1, & y_{h_j^{(r)}} = y_{v_i^{(f)}} \\ 0, & y_{h_j^{(r)}} \neq y_{v_i^{(f)}}, \end{cases} \end{aligned} \tag{5.14}$$

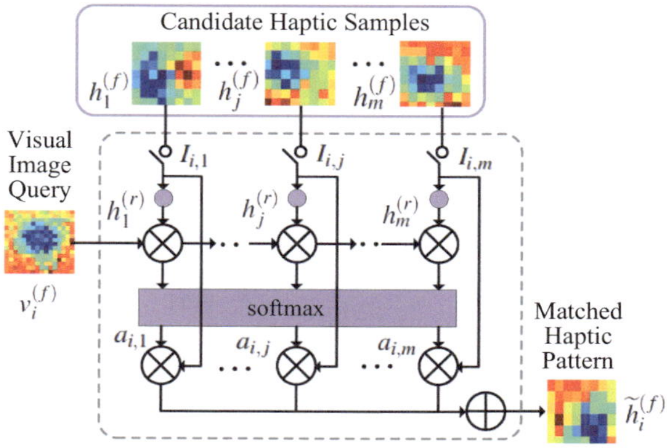

Fig. 5.6 Attention-based haptic-visual data matching process

where $y_{h_j^{(r)}}$ and $y_{v_i^{(f)}}$ respectively denote category labels of the candidate haptic representation $h_j^{(r)}$ and visual query vector $v_i^{(f)}$. By this way, haptic samples having the same category labels with the query visual vector are selected as candidates.

Subsequently, $Sim_{i,j}$ is normalized to the attention weights $a_{i,j}$ by the softmax function, reflecting the matching importance of the candidate haptic samples. Finally, by summation of the candidate haptic feature vectors with their attention weights, the haptic pattern $\widetilde{h}_i^{(f)}$ is obtained, which is strongly paired with $v_i^{(f)}$,

$$\begin{aligned} a_{i,j} &= Softmax\left(Sim_{i,j}\right), \\ \widetilde{h}_i^{(f)} &= \sum_{j=1}^{m} a_{i,j} \cdot h_j^{(f)}. \end{aligned} \tag{5.15}$$

In summary, the implementation of the attention network results in a significant differentiation in the attention weights assigned to haptic samples relative to the query visual vector. Haptic samples that exhibit a high degree of relevance to the query visual vector are allocated larger attention weights, whereas those with lesser relevance receive smaller weights. Consequently, the attention network effectively constructs an optimal haptic pattern that is specifically tailored to each query visual image. This mechanism ensures that the resulting haptic pattern is strongly paired with the query visual image on a one-to-one basis. As a result, the issue of unpaired training data, which is prevalent in cross-modal communications, can be effectively addressed.

Strongly paired semantic association

After employing haptic-visual modality matching, the next step is to construct the strongly paired semantic association. Firstly, the intra-modal discrimination for the derived haptic patterns should be concerned. For the haptic modality, it is essential to minimize the variation among the obtained haptic patterns that belong to the same category, while maximizing the variation among patterns from different categories. This method is analogous to the scheme described in Sect. 5.2.1, where a classifier implemented using a softmax layer in conjunction with an adversarial principle is employed. Given the predicted category labels, denoted by $\widetilde{h}^{(c)}$, and the real category labels y, the intra-modal discrimination loss, parameterized by $\boldsymbol{\theta}_H$, is defined as

$$\min_{\boldsymbol{\theta}_H} \max_{\boldsymbol{\beta}} L_{cat}^{H}\left(D_2, \widetilde{h}^{(c)}\right) = \mathbb{E}_{\widetilde{h}^{(c)}}\left[\log\left(1 - D_2\left(\widetilde{h}^{(c)}\right)\right)\right] + \mathbb{E}_{\mathbf{y}}\left[\log D_2(y)\right], \tag{5.16}$$

where D_2 denotes the discriminator with parameters $\boldsymbol{\beta}$.

Besides intra-modal discrimination, the inter-modal semantic consistency for the derived visual-haptic pairs$\{(v_1^{(f)}, \widetilde{h}_1^{(f)}), \ldots, (v_i^{(f)}, \widetilde{h}_i^{(f)}), \ldots, (v_n^{(f)}, \widetilde{h}_n^{(f)})\}$ should also be considered. To ensure inter-modal semantic consistency, the distances between strongly paired haptic patterns and visual image features should be minimized, while the distances between mismatched pairs should be maximized, albeit not exceeding a predefined threshold. Based on this, the semantic loss, parameterized by w and b, is

$$L_{cro} = \sum_{i=1}^{n} \left\| \widetilde{h}_i^{(f)} - v_i^{(f)} \right\|_2^2 + \sum_{i \neq j} \max\left(\gamma - \left\| \widetilde{h}_i^{(f)} - v_j^{(f)} \right\|, 0\right)^2, \tag{5.17}$$

where the threshold value $\gamma = 1$. Through the meticulous design of the aforementioned functions, a fine-grained semantic association has been successfully established, facilitated by the strongly paired haptic-visual signals. Specifically, upon the determination of the parameters via the optimization strategy detailed in Sect. 5.2.4, the resultant haptic patterns are imbued with fine-grained semantic information that transcends modality-specific characteristics.

5.2.3 High-Level Global-Local Image Synthesis

Upon the establishment of the middle-level fine-grained semantic association, the subsequent and final module involves the realization of cross-modal image synthesis. Traditional methods are predominantly designed based on the MSE or the L_2 norm loss metric. These approaches necessitate that the generated images achieve precise pixel-level correspondence with their respective real images. However, such metrics fail to account for global information, often resulting in generated images that appear blurry or lacking in detail. In recent years, GAN, taken as a representative generation method, has demonstrated potential in bridging modalities [14]. Therefore, GAN is adopted here to perform haptic-to-image synthesis. Meanwhile, an adversarial loss function is employed to capture the global structure of images, leveraging the fine-grained semantic information embedded in the haptic patterns. In other words, it can achieve a balance between local pixel-level accuracy and global structural coherence,

$$\begin{aligned} \min_{\eta} \max_{\gamma} L_{\text{gen}}(G, D_3) = \lambda_1 \mathbb{E}_v \left[\log D_3\left(v, G\left(\widetilde{h}^{(f)}\right)\right)\right] + \mathbb{E}_{(v, v_{gen})} \left[\left\| v - G\left(\widetilde{h}^{(f)}\right) \right\|_2\right] \\ + \lambda_1 \mathbb{E}_{v_{\text{gen}}} \left[\log\left(1 - D_3\left(v, G\left(\widetilde{h}^{(f)}\right)\right)\right)\right]. \end{aligned} \tag{5.18}$$

Specifically, the generator G parameterized by η tries to minimize the second and third items in Eq. (5.18). Simultaneously, the discriminator D_3, parameterized by $\boldsymbol{\gamma}$, seeks to maximize the distinction between real image v and the synthesized image $v_{\text{gen}} \sim G(\widetilde{h}^{(f)})$ in the first and second items. λ_1 serves as a hyperparameter to balance the relative contributions of the local L_2 norm loss and the global adversarial loss.

Finally, the joint objective function, which integrates the middle-level fine-grained semantic association and the high-level global-local image synthesis, is formulated

$$\min_{\boldsymbol{\theta}_H,\boldsymbol{\eta},(w,b)\boldsymbol{\beta},\boldsymbol{\gamma}} \max L' = \lambda_2 L_{cat}^T\left(D_2, \widetilde{h}^{(c)}\right) + L_{cro} + L_{gen}\left(G, D_3\right), \tag{5.19}$$

where λ_2 is also a hyperparameter.

5.2.4 *Optimization Process*

Based on the preceding descriptions about the FHIS, the optimization process during training can be decomposed into two alternating stages. These stages iteratively update the parameters by minimizing the objective functions defined in Eq. (5.12) and Eqs. (5.16)–(5.19). The SGD algorithm is employed to iteratively obtain the parameter sets $\{\boldsymbol{\theta}_V, \boldsymbol{\alpha}\}$ and $\{\boldsymbol{\theta}_H, \boldsymbol{\eta}, (w, b), \boldsymbol{\beta}, \boldsymbol{\gamma}\}$ until convergence is achieved. The optimization process is summarized in Algorithm 5.2.

Upon completion of the optimization process, the FHIS approach can be employed in scenarios where new haptic packets are received at the master terminal, but the corresponding desired visual packets either fail to arrive, are impaired, or are lost within a specified time duration. Specifically, the newly received haptic signal, decoded from the haptic packet h' is firstly processed through the pre-trained GRU network, as well as the trained fully connected layers and attention network depicted in Fig. 5.5. The resultant haptic pattern, which encapsulates semantic information, is then utilized as the input to the trained generator G. The output of the generator G is the synthesized visual image v'_{gen}, which corresponds to the input haptic signal. It is important to note that for cases where the visual packet is merely delayed rather than lost, Algorithm 5.2 can be executed to fine-tune the parameters once the delayed packet arrives. This fine-tuning process further enhances the effectiveness and accuracy of the FHIS approach in subsequent applications.

5.2.5 *Experimental Result and Analysis*

In this section, the FHIS approach is employed on a standard dataset. Its performance can be evaluated by comparing with the other competing approaches.

Experimental configuration

Dataset: The LMT surface material database [6] is employed here, which is the same as that in Sect. 5.1.5. Specifically, each material surface is associated with ten pairs of visual images and haptic samples, with their respective coarse categories explicitly provided. To emulate the unpaired conditions typically encountered at the receiver end of a cross-modal communication system, the visual images and haptic samples are intentionally disordered. Furthermore, to ensure an adequate volume of data for training purposes, the visual images are augmented to a total of one hundred

Algorithm 5.2 Optimization process of the FHIS [10]

Require: Visual dataset $V = \{v_1, \ldots, v_i, \ldots, v_n\}$, haptic dataset $H = \left\{h_1, \ldots, h_j, \ldots, h_m\right\}$ with category labels; Batch-size, learning rate μ_1, μ_2, iterations n_1, n_2

```
Initialize parameters: {θ_V, α}, {θ_H, η, (w, b), β, γ},
Set hyper-parameters: λ_1, λ_2 and learning rate: μ_1, μ_2,
Set mini-batch size and iteration number epoch.
for epoch do
   Update θ_V and α from Eq. (5.12),
   α ← α − μ_1 ∇_α L^V_cat (D_1, v^(c)),   { update D_1},
   θ_V ← θ_V − μ_1 ∇_{θ_V} L^V_cat (D_1, v^(c));
   Update β from Eq. (5.16),
   β ← β − μ_2 ∇_β L^T_cat (D_2, h̃^(c));   {update D_2}
   Update γ from Eq. (5.18),
   γ ← γ − μ_2 ∇_γ L_gen (G, D_3);   {update D_3}
   Update θ_H, w, b, η from Eq. (5.19),
   (θ_H, w, b) ← (θ_H, w, b) − μ_2 ∇_(θ_H,w,b) L',
   η ← η − μ_2 ∇_η L'.   {update G }
end for
```

Output: Parameter set in the FHIS: $\{\boldsymbol{\theta}_V, \boldsymbol{\alpha}\}, \{\boldsymbol{\theta}_H, \boldsymbol{\eta}, (w, b), \boldsymbol{\beta}, \boldsymbol{\gamma}\}$

through a series of geometric transformations, including translation and rotation. In the final dataset preparation, 80% of the visual images and haptic samples are randomly selected for training, while the remaining 20% were reserved for testing and performance evaluation. It is important to note that this database does not provide explicit one-to-one correspondence between visual and haptic samples, as these modalities were collected independently.

Competing schemes: Here, five competing schemes are selected, including DCCAE [15], CGAN [16], ACGAN [17], AE-GAN [18], and No-Attention. Specifically, the deep canonically correlated autoencoder (DCCAE) comprises dual autoencoders and is trained using constraints derived from canonical correlation analysis and recovery error. It is applicable for cross-modal image synthesis tasks. The conditional generative adversarial network (CGAN) is employed to produce images based on the semantic information extracted from labeled haptic signals, serving as a comparison benchmark. The auxiliary classifier generative adversarial network (ACGAN) is a conditional image synthesis technique that incorporates an auxiliary classifier. Within the ACGAN, the classifier receives inputs that are a composite of noise and haptic features, while the discriminator provides manual annotations for the input data. The autoencoder generative adversarial network (AE-GAN) integrates an autoencoder with a GAN. When adapted for haptic-to-image synthesis, the autoencoder component remains intact, and the generator's input consists of a combination of noise and haptic features. The No-Attention variant refers to the FHIS model without the attention network.

Implementation details: The implementation details of all the modules and networks in the FHIS are listed in the following.

Feature extraction module: It is constructed using a three-layer CNN, followed by a three-layer fully connected network with node configurations of 1024, 128, and

k, respectively, and culminates in a softmax layer with C nodes. The input visual images for training are standardized to a size of $256 \times 256 \times 3$. The CNN employs convolutional kernels of size 3×3, with the number of nodes in each layer being 64, 128, and 256, respectively. In the full-connected network, the number of nodes in the last layer, denoted as k, represents the dimensionality of the extracted visual features. Additionally, the discriminator D_1 is composed of three fully connected layers with node configurations of 512, 256, and 1.

Semantic association module: It is designed to extract haptic features using a gated recurrent unit (GRU) network and a three-layer feedforward neural network with node configurations of 1024, 128, and 8. The architecture of the attention network is depicted in Fig. 5.6. Furthermore, the discriminator D_2 shares a similar structure with D_1.

Image synthesis module: It employs a generator G that is constructed using a four-layer deconvolutional network. Within G, the convolutional kernels have a size of 3×3, and the number of nodes in the four layers are 256, 128, 64, and 3, respectively. The discriminator D_3 is designed to be analogous to D_2.

Results and analysis

Firstly, to comprehensively evaluate the performance of the FHIS approach, both qualitative and quantitative comparisons with other competing methods are presented. When concerning qualitative comparison, Fig. 5.7 illustrates several visual images recovered by the aforementioned approaches, including FHIS, CGAN, ACGAN, and AE-GAN. A visual inspection reveals that the FHIS approach produces images with color and contour that closely resemble the original real images, and the textures are also distinctly clear. In contrast, the CGAN, ACGAN, and AE-GAN methods, which are variants of GAN, require weakly paired training data, where visual images and haptic samples share corresponding category labels. However, such weakly paired data may not be guaranteed in cross-modal communication systems, leading to inconsistent correspondences between modalities. The proposed FHIS approach, on the other hand, is capable of mining strongly paired relationships between haptic and visual modalities during the training process. This capability effectively addresses the issue of inconsistent correspondence among different modality signals, ultimately achieving superior image synthesis results.

The quantitative evaluation focuses on determining the accuracy of the category classification for the generated images. The ground truth for this evaluation is derived from the category labels of the test visual images. Three metrics are employed for this assessment: MicroF1, MacroF1, and Accuracy-SVM (ACC-SVM). The ACC-SVM metric utilizes a pre-trained linear support vector machine (SVM) as the standard classifier to identify the categories of the synthesized images. The result is summarized in Table 5.4. It indicates that the FHIS approach significantly outperforms the other competing methods. Specifically, the DCCAE approach exhibits the lowest performance among all methods. The AE-GAN and ACGAN methods achieve slightly higher accuracy than CGAN, likely due to the incorporation of semantic classifiers in their image synthesis processes. However, these improvements are still insufficient to match the performance of FHIS. It is noteworthy that the average accu-

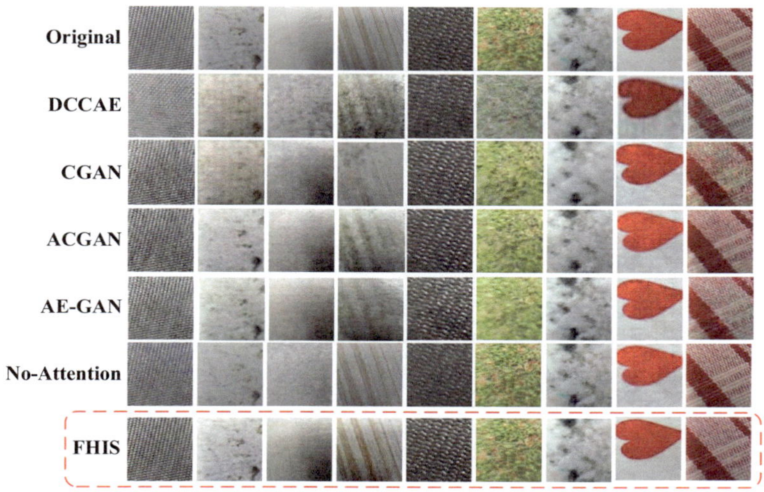

Fig. 5.7 The recovered visual images from haptic signals by the FHIS and the other competing approaches

Table 5.4 Quantitative result of the FHIS and the other competing approaches

Approach	MiroF1	MacroF1	ACC-SVM
DCCAE	0.544	0.449	0.431
CGAN	0.568	0.451	0.609
ACGAN	0.571	0.461	0.656
AE-GAN	0.575	0.455	0.654
No-attention	0.574	0.462	0.667
FHIS	**0.588**	**0.476**	**0.743**

racy of ACC-SVM is 72.4%. This relatively low accuracy is attributed to the high similarity between haptic signals in the "glossy" and "foils and papers" categories, which makes them extremely difficult to distinguish. This similarity, in turn, affects the discriminability of the generated images by the FHIS approach. For the remaining seven categories in the database, the ACC-SVM accuracy is approximately 90%, demonstrating that the FHIS approach is highly effective for most material categories.

Next, the effectiveness of the FHIS approach is rigorously evaluated. Figure 5.8 presents visual images synthesized by the FHIS approach across different categories. A visual inspection reveals that the generated images exhibit rich and distinct semantic features, accurately capturing fine-grained texture information within each category. This demonstrates the FHIS approach's ability to generate high-quality and semantically meaningful images. Additionally, the contributions of specific modules within the FHIS framework are analyzed by comparing the full FHIS approach with three degraded variants. The first degraded variant removes the inter-modal consistency design principle from the FHIS approach, denoted as

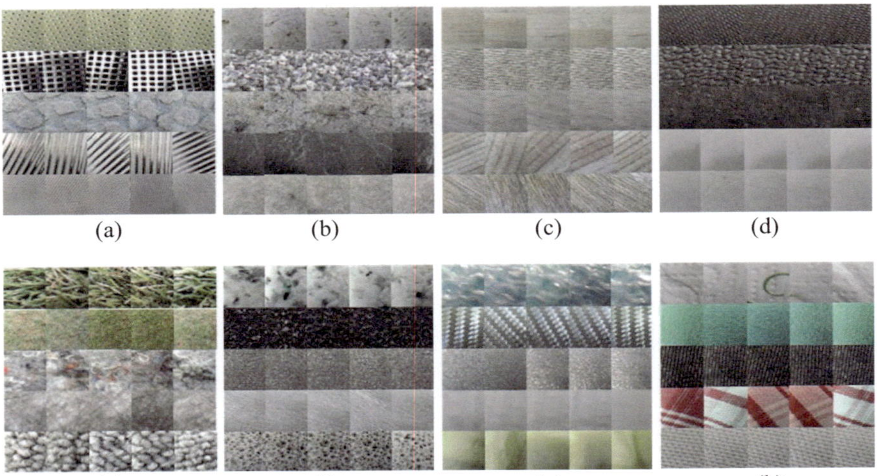

Fig. 5.8 The generated visual images belonging to different categories by the FHIS: **a** Mesh; **b** Stone; **c** Wood; **d** Rubber; **e** Fiber; **f** Foam; **g** Foil and paper; **h** Textiles and fabrics

"Only intra-modal", focusing solely on intra-modal discrimination within the haptic modality. This scheme aims to verify the contribution of the "strongly paired semantic association" mechanism described in Sect. 5.2.2. The second degraded variant, denoted as "Weakly paired", replaces the strongly paired haptic-visual data matching based on the attention network with weakly paired data, which only considers the consistency of coarse category labels. This scheme evaluates the effectiveness of the "attention-based haptic-visual data matching" mechanism described in Sect. 5.2.2. The third degraded variant, denoted as "Only MSE", retains only the MSE loss, which measures the Euclidean distance between the generated image and the real image. This scheme assesses the combined effect of the "global" adversarial loss and the "local" MSE loss described in Sect. 5.2.3.

Here, the inception score (IS) is employed as a quantitative metric to evaluate the quality and category richness of the generated images v_{gen} . The IS is calculated as follows:

$$IS = \exp\left(\mathbb{E}_{v_{gen}} KL\left(p\left(y_{gen} \mid v_{gen}\right) \| p\left(y_{gen}\right)\right)\right) \tag{5.20}$$

where y_{gen} represents the label predicted by the inception model. The underlying principle of this metric is that effective approaches should generate diverse yet meaningful images. Therefore, a larger Kullback-Leibler (KL) divergence between the marginal distribution $p(y_{gen})$ and the conditional distribution $p(y_{gen} \mid v_{gen})$ indicates better performance.

Table 5.5 presents the mean and standard deviation of the IS for real images, the FHIS approach, and the three degraded variants. The result indicates that the images

Table 5.5 IS about real images, images recovered by the FHIS, and those by the three degraded variants

Approach	Inception score
Real images	10.047 ± 1.088
FHIS	$\mathbf{9.684 \pm 1.102}$
Only intra-modal	8.618 ± 0.876
Weakly paired	8.064 ± 1.171
Only MSE	9.134 ± 0.874

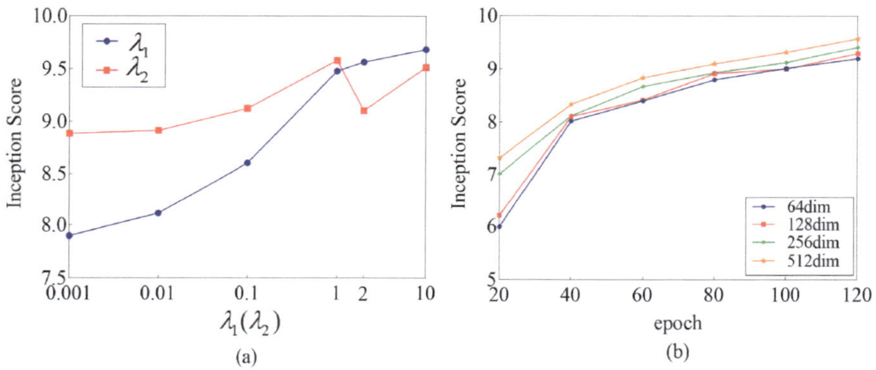

Fig. 5.9 Inception score versus **a** λ_1 and λ_2; **b** k and epoch

generated by the FHIS approach are closest to the real images and outperform the other three variants. This highlights the significance of the components designed in Sects. 5.2.2 and 5.2.3. Notably, the weakly paired schemes exhibit relatively poor performance, demonstrating that they are insufficient for achieving high-quality cross-modal image synthesis.

Finally, the sensitivity of key hyperparameters and architectural choices in the FHIS approach is analyzed. Specifically, the impacts of the hyperparameters λ_1 and λ_2, the dimension of features k, and the number of training epochs are examined. λ_1 and λ_2 regulate the adversarial loss and the haptic intra-modal discrimination loss, respectively. To visualize their influence, the IS of the generated images is plotted against varying values of these hyperparameters. Figure 5.9a illustrates the impacts of these two hyperparameters. The result shows that as λ_1 increases, the IS initially rises and eventually stabilizes. As λ_2 increases, the IS exhibits slight fluctuations. This verifies the critical roles of two additional discriminators D_2, D_3, providing guidance about the hyperparameter setting. Figure 5.9b presents the sensitivity analysis of the number of training epochs and the feature dimension k. The IS is evaluated under different settings to assess the convergence behaviors. These findings are essential for fine-tuning the FHIS approach to achieve the best possible performance in haptic-to-image synthesis task.

5.3 Visual-Haptic Mutual Signal Recovery

In contrast to haptic-dominant and audio-visual-dominant multi-modal services, all modalities carry equal importance in the third category of services, that is equally important multi-modal services. This may present two primary challenges in the design of approaches at the receiver. On one hand, the diminished emphasis on any single modality intensifies competition for limited transmission and processing resources. For example, in remote practical teaching, the quality of both audio-visual learning materials and haptic feedback significantly impacts students' engagement and learning outcomes. Consequently, it is crucial to simultaneously meet high quality of all the received modality signals. On the other hand, all modalities are equally susceptible to noise and interference. Existing schemes assume that one modality signal can be received intact, which will be utilized for recovering the other modality signal. For instance, the AVHR synthesizes a "virtual" touch sensation based on accurately received visual signals [1]. The FHIS restores the impaired visual images with the assistance of complete haptic streaming [10]. However, this assumption is often untenable in the equally important multi-modal services, where ensuring flawless transmission of any single modality becomes increasingly challenging due to various factors such as network congestion, signal interference, and limited bandwidth. Moreover, existing cross-modal signal recovery schemes rely on the utilization of potential correlations or semantics among modalities. These semantics can be derived from an external database [1] or extracted from annotated internal semantic labels [10], neglecting the unique characteristics of visual and haptic modalities.

To address the aforementioned challenges, this section proposes a visual-haptic mutual signal recovery (VHMR) scheme for the damaged, lost, or delayed signals [19]. It aims to relax the assumption that flawless reception of one modality is required to ensure the integrity and quality of the overall multi-modal experience. Figure 5.10 illustrates the framework of the VHMR. It is underpinned by two core mechanisms: modality-aware knowledge transfer and semantic-aware signal generation. For the former, haptic and visual knowledge are respectively extracted from the related external databases, focusing on their unique characteristics. For the latter, both coarse-grained and fine-grained signal recovery are carefully and respectively designed by leveraging intra-modal and inter-modal semantics. Therefore, it contains four modules: feature extraction by knowledge transfer, semantic association by modality fusion, coarse-grained signal recovery by adversarial learning, and fine-grained signal recovery by cyclic-consistency constraint.

5.3.1 *Feature Extraction by Knowledge Transfer*

The first module is to extract features from visual and haptic signals. However, in real-world situations, the amount of the received visual and haptic signals is often insufficient. To solve this problem, the auxiliary visual and haptic databases stored in

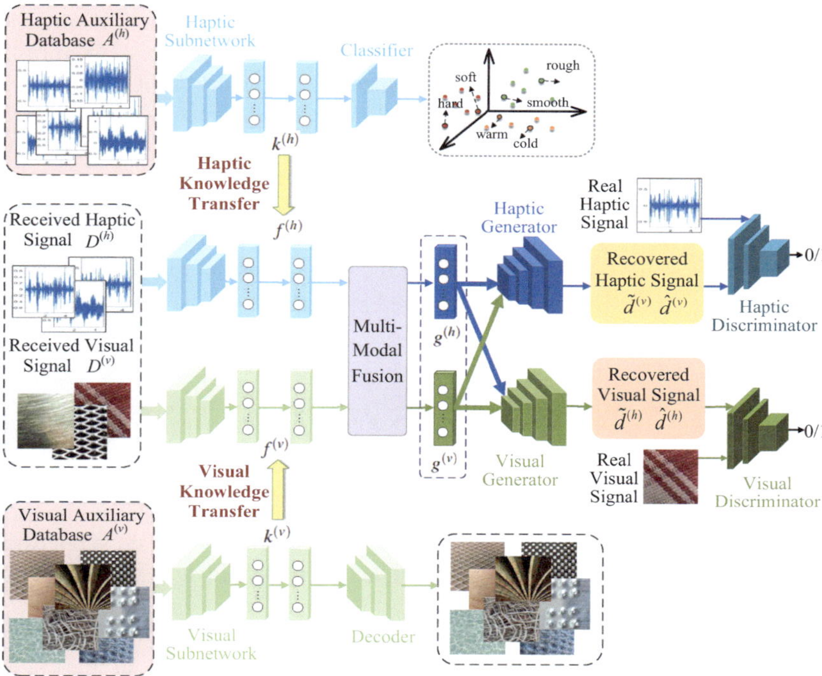

Fig. 5.10 Framework of the VHMR

the cloud can be used to provide extra knowledge. Furthermore, for visual modality, many large-scale auxiliary visual databases lack meaningful labels due to the absence of manual annotation, which affects the quality of knowledge formation. For haptic modality, there are no large-scale haptic databases available for knowledge formation. Moreover, effectively transferring knowledge to drive feature extraction is a crucial issue that needs to be addressed.

In order to realized knowledge formation, we introduce two auxiliary databases: the visual database $A^{(v)} = \{a_j^{(v)}\}_{j=1}^{N_1}$ and the haptic database $A^{(h)} = \{a_j^{(h)}, z_j^{(h)}\}_{j=1}^{N_2}$. For visual modality, an autoencoder is employed to extract knowledge in an unsupervised manner. Specifically, a sample $a^{(v)}$ is fed into the visual subnetwork and several fully connected layers to generate visual knowledge $k^{(v)}$, where $k^{(v)} = KF(a^{(v)}; \theta_{kf}^{(v)})$. Subsequently, $k^{(v)}$ is passed through a reverse structure-based neural network to reconstruct the visual sample $\tilde{a}^{(v)}$, that is $\tilde{a}^{(v)} = RC(k^{(v)}; \theta_{rc}^{(v)})$. $\theta_{kf}^{(v)}$ and $\theta_{rc}^{(v)}$ represent the related parameters. Based on this, the reconstruction loss is

$$\mathcal{L}_{AV} = \sum_{j=1}^{N_1} \left\| \tilde{a}_j^{(v)} - a_j^{(v)} \right\|_2^2. \tag{5.21}$$

In contrast to visual signals, the variety of typical haptic signals is often less extensive. For example, when characterizing the surface of a material, five primary haptic attributes, such as roughness and hardness, are typically sufficient to provide an effective description [7]. Therefore, a small-scale haptic database can be utilized for the acquisition of haptic knowledge. This database can be annotated with properties such as roughness, hardness, and others, thereby facilitating the extraction of specific haptic information. Specifically, the haptic input $a^{(h)}$ is passed through the haptic subnetwork and several fully connected layers to obtain knowledge $k^{(h)}$, $k^{(h)} = KF(a^{(h)}; \theta_{kf}^{(h)})$. Subsequently, $k^{(h)}$ is fed into a classifier to determine the relevant properties. For utilizing the real tagged properties $z^{(h)}$, the classification loss is designed as

$$\mathcal{L}_{AH} = -\frac{1}{N_2}\sum_{j=1}^{N_2} z_j^{(h)}\left[\log p\left(k_j^{(h)}; \theta_{rc}^{(h)}\right)\right], \tag{5.22}$$

where $p(\cdot)$ is the classifier with parameter $\theta_{rc}^{(h)}$. $z^{(h)}$ denotes various combinations of haptic properties. By minimizing $\mathcal{L}_{AV}$ and $\mathcal{L}_{AH}$, the visual/haptic subnetworks and full-connected layers can be determined, and the knowledge $k^{(v)}$ and $k^{(h)}$ can be derived.

Upon the acquisition of knowledge, it necessitates the subsequent transfer to the feature extraction network situated at the receiver's edge node. In this context, it is presumed that the edge node encompasses M instances of paired visual-haptic signal data for training purposes, specifically denoted as $D^{(v)} = \{d_i^{(v)}\}_{i=1}^M$ and $D^{(h)} = \{d_i^{(h)}\}_{i=1}^M$. Herein, $d_i^{(v)} \in \mathbb{R}^v$ represents the visual signal, while $d_i^{(h)} \in \mathbb{R}^h$ signifies the haptic signal. For visual modality, the feature extraction network comprises a visual subnetwork and several fully connected layers, which are employed to extract visual features $f^{(v)}$ from $d^{(v)}$, expressed as $f^{(v)} = FE(d^{(v)}; \theta_{fe}^{(v)})$. Analogously, for haptic modality, the feature extraction process is represented by $f^{(h)} = FE(d^{(h)}; \theta_{fe}^{(h)})$.

Specifically, the visual and haptic subnetworks at the edge nodes are capable of sharing the parameters from the cloud. To further enhance the efficacy of knowledge transfer, it is imperative to minimize the maximum mean discrepancy (MMD) between $A^{(v)}$, $A^{(h)}$ and $D^{(v)}$, $D^{(h)}$, as indicated in [20]. In light of this consideration, the loss function for knowledge transfer is formulated as

$$\begin{aligned}\mathcal{L}_{\text{tran}} &= \left\|\mathbb{E}_{A^{(v)}}\left[\phi\left(k^{(v)}\right)\right] - \mathbb{E}_{D^{(v)}}\left[\phi\left(f^{(v)}\right)\right]\right\|_{\mathcal{H}}^2 \\ &+ \left\|\mathbb{E}_{A^{(h)}}\left[\phi\left(k^{(h)}\right)\right] - \mathbb{E}_{D^{(h)}}\left[\phi\left(f^{(h)}\right)\right]\right\|_{\mathcal{H}^2}^2 .\end{aligned} \tag{5.23}$$

Here, $\phi(\cdot)$ denotes the mapping function that projects the variables into the reproducing kernel Hilbert space. By minimizing $\mathcal{L}_{\text{tran}}$, the feature extraction networks at the edge node can be aligned with those of the auxiliary databases. This enables the determination of $\theta_{fe}^{(v)}$ and $\theta_{fe}^{(h)}$, thereby yielding $f^{(v)}$ and $f^{(h)}$.

5.3.2 Semantic Association by Modality Fusion

Following the derivation of $f^{(v)}$ and $f^{(h)}$, the next module involves establishing a semantic association between the visual and haptic modalities. To achieve this, $f^{(v)}$ and $f^{(h)}$ are mapped into a shared semantic space while adhering to two critical principles. Firstly, the semantic gap between visual and haptic features belonging to the same content should be minimized. Secondly, within each modality (visual and haptic), the discriminative ability for features corresponding to different categories should be preserved.

Concretely, by mapping $f^{(v)}$ and $f^{(h)}$ into the shared semantic space, representations $g^{(v)}$ and $g^{(h)}$ are respectively obtained, $g^{(v)} = CM(f^{(v)}; \theta_{cm}^{(v)})$, $g^{(h)} = CM(f^{(h)}; \theta_{cm}^{(h)})$. Here, $CM(.,.)$ is the mapping function, while $\theta_{cm}^{(v)}$ and $\theta_{cm}^{(h)}$ are parameters of the supported multi-layer neural network. To eliminate modality-specific differences, the distance between $g^{(v)}$ and $g^{(h)}$, (i.e. those corresponding to the same content) should be minimized. Conversely, the distance between the unpaired $g^{(v)}$ and $g^{(h)}$ should be maximized. It should be noted that the "paired" instances refer to the matched visual and haptic signals prior to transmission, which can be identified by the time stamps embedded in the respective packets. Based on this, visual and haptic signals with identical time stamps in their corresponding packets are considered "paired", whereas those with different time stamps are deemed "unpaired". The correlation loss function can be designed accordingly as

$$\mathfrak{L}_{cor} = \sum_{i=1}^{M} \left\| g_i^{(v)} - g_i^{(h)} \right\|_2^2 + \sum_{i \neq j} \max\left(\gamma - \left\| g_i^{(v)} - g_j^{(h)} \right\|_1, 0\right)^2, \tag{5.24}$$

where γ is the upper bound of the unpaired feature distance.

To ensure the preservation of discriminative characteristics within the shared semantic space, it is imperative that $g^{(v)}$ or $g^{(h)}$ belonging to the same category exhibit greater proximity than those from distinct categories. This objective can be achieve through the implementation of two additional classification tasks for $g^{(v)}$ and $g^{(h)}$, respectively, under the guidance of category labels. The discrimination loss function, constructed based on the multi-class cross-entropy principle, is defined as

$$\mathfrak{L}_{cat} = -\frac{1}{M} \sum_{i=1}^{M} \left[y_i^{(v)} \log p\left(g_i^{(v)}, \theta_{cl}^{(v)}\right) + y_i^{(h)} \log p\left(g_i^{(h)}, \theta_{cl}^{(h)}\right) \right], \tag{5.25}$$

where $p(\cdot)$ represents the classifiers with parameters $\theta_{cl}^{(v)}$ and $\theta_{cl}^{(h)}$. $y^{(v)}$ and $y^{(h)}$ denote the corresponding real category labels (e.g. "mesh", "stone", "metal", etc.), which are encoded as one-hot vectors.

Consequently, the overall objective function for semantic association is formulated by integrating the aforementioned two loss functions, thereby ensuring the efficacy of semantic association across modalities, $\mathfrak{L}_{cor} + \alpha \mathfrak{L}_{cat}$. By minimizing this objective function, semantic similarity among the paired visual and haptic features can be effectively extracted. Finally, the optimal parameters $\theta_{cm}^{(v)}, \theta_{cm}^{(h)}, \theta_{cl}^{(v)}, \theta_{cl}^{(h)}$, are identified, and $g^{(v)}$, $g^{(h)}$ can be got.

5.3.3 Coarse-Grained Signal Recovery by Adversarial Learning

Upon the completion of semantic association, the recovery of visual and haptic signals can be achieved based on the representations $g^{(v)}$ and $g^{(h)}$. In this context, a GAN is employed as the primary instrument for the coarse-grained signal recovery module. Taking the visual modality as an exemplar, within the GAN framework, the generator GV transforms $g^{(v)}$ into the recovered visual signals $\widetilde{d}^{(v)}$, as expressed by the equation $\widetilde{d}^{(v)} = GV(g^{(v)}; \theta_{rec}^{(v)})$. Concurrently, the discriminator DV is tasked with differentiating between the real visual signal $d^{(v)}$ and the recovered signal $\widetilde{d}^{(v)}$.

Specifically, the generator GV aims to produce $\widetilde{d}^{(v)}$ in such a manner as to deceive the discriminator DV, while the discriminator DV endeavors to enhance its discriminative capabilities to distinguish $\widetilde{d}^{(v)}$ from $d^{(v)}$. The objective function can be formed as

$$\begin{aligned}
\mathcal{L}_{DV} &= - \mathbb{E}_{d^{(v)}} \left[\log DV \left(d^{(v)}; \theta_{dis}^{(v)} \right) \right] - \mathbb{E}_{\widetilde{d}^{(v)}} \left[\log \left(1 - DV \left(\widetilde{d}^{(v)}; \theta_{dis}^{(v)} \right) \right) \right], \\
\mathcal{L}_{GV} &= - \mathbb{E}_{\widetilde{d}^{(v)}} \left[\log DV \left(\widetilde{d}^{(v)}; \theta_{dis}^{(v)} \right) \right],
\end{aligned} \tag{5.26}$$

where $\theta_{dis}^{(v)}$ and $\theta_{rec}^{(v)}$ respectively denotes parameter sets of DV and GV. By minimizing $\mathcal{L}_{DV}$ and $\mathcal{L}_{GV}$, GV is able to approximate the distribution of $d^{(v)}$ and utilize this approximation to guide the generation of $\widetilde{d}^{(v)}$.

In addition to the incorporation of adversarial loss, the pixel-wise loss is also considered to further align $\widetilde{d}^{(v)}$ with $d^{(v)}$, which is expressed as

$$\mathcal{L}_{PV} = \frac{1}{M} \sum_{i=1}^{M} \left\| d_i^{(v)} - \widetilde{d}_i^{(v)} \right\|_2^2. \tag{5.27}$$

Consequently, the comprehensive objective function for the coarse-grained visual signal recovery is formulated as $\mathcal{L}_{GV} + \beta_V \mathcal{L}_{PV}$.

Analogously, for the coarse-grained haptic signal recovery following the acquisition of $g^{(h)}$, the combination of adversarial loss via GAN and point-to-point loss is also employed. The generator GH converts $g^{(h)}$ into the recovered haptic signal $\widetilde{d}^{(h)}$, while discriminator DH differentiates between the real haptic signal $d^{(h)}$ and $\widetilde{d}^{(h)}$. The corresponding loss functions for the haptic modality, denoted as $\{\mathcal{L}_{DH}, \mathcal{L}_{GH}\}$ and $\mathcal{L}_{PH}$, are structured in a manner analogous to Eqs. (5.26) and (5.27), with parameters $\theta_{dis}^{(h)}$ and $\theta_{rec}^{(h)}$. By minimizing the losses $\mathcal{L}_{DV}, \mathcal{L}_{GV} + \beta_V \mathcal{L}_{PV}$ and $\mathcal{L}_{DH}, \mathcal{L}_{GH} + \beta_H \mathcal{L}_{PH}$, the respective parameters can be optimized, thereby enabling the generation of the recovered visual and haptic signals $\widetilde{d}^{(v)}$ and $\widetilde{d}^{(h)}$.

5.3.4 Fine-Grained Signal Recovery by Cyclic-Consistency Constraint

Despite the semantic association across visual and haptic modalities being performed and utilized as input, the training of GV and GH remains intra-modal, as detailed in Sect. 5.3.3. This implies that the inter-modal semantics are not fully exploited. To further enhance the quality of the recovered signals, we propose to achieve a fine-grained signal recovery for each modality by incorporating a cyclic-consistency constraint.

Specifically, given that the dimensions of the two modalities are identical in the shared semantic space, the visual representation $g^{(v)}$, derived from $d^{(v)}$, is input into the trained haptic generator GH in Sect. 5.3.3. Consequently, the recovered haptic signal $\widehat{d}^{(h)}$ is obtained, expressed as $\widehat{d}^{(h)} = GH(g^{(v)}; \theta_{\text{rec}}^{(h)})$. $\widehat{d}^{(h)}$ is then fed back into the haptic feature extraction module (Sect. 5.3.1) and the semantic association module (Sect. 5.3.2), yielding the haptic representation $\widehat{g}^{(h)}$. To enforce semantic consistency, the cosine distance is employed to constrain $\widehat{g}^{(h)}$ and the corresponding $g^{(v)}$ in the shared semantic space, defined as follows:

$$\mathcal{L}_{cycH} = \frac{g^{(v)} \cdot \widehat{g}^{(h)}}{\left\|g^{(v)}\right\|_2 \cdot \left\|\widehat{g}^{(h)}\right\|_2} = \frac{\sum_{i=1}^{M} g_i^{(v)} \widehat{g}_i^{(h)}}{\sqrt{\sum_{i=1}^{M} \left(g_i^{(v)}\right)^2} \cdot \sqrt{\sum_{i=1}^{M} \left(\widehat{g}_i^{(h)}\right)^2}}. \tag{5.28}$$

Similarly, the haptic representation $g^{(h)}$ is input into the trained visual generator GV, $\widehat{d}^{(v)} = GV(g^{(h)}; \theta_{rec}^{(v)})$. The recovered visual signal $\widehat{d}^{(v)}$ is fed back into the visual feature extraction and semantic association modules to get $\widehat{g}^{(v)}$. The cosine distance $\mathcal{L}_{cycV}$ is likewise employed to constrain $\widehat{g}^{(v)}$ and the corresponding $g^{(h)}$.

The inter-modal cyclic information flow described above is depicted in Fig. 5.11. The final loss function is defined as $\lambda_V \mathcal{L}_{cycV} + \lambda_H \mathcal{L}_{cycH}$. λ_V and λ_H are hyperparameters that control the weights of the visual and haptic modalities, respectively. By minimizing this final loss function, the optimal parameters of GV and GH can be obtained.

As demonstrated in the Sect. 5.3.6, this approach effectively enhances the recovery capabilities of the generators for both modalities, thereby realizing fine-grained signal recovery through the cyclic-consistency constraint.

5.3.5 Optimization Process

The optimization process of the VHMR can be categorized into three primary stages, as illustrated in Algorithm 5.3. Upon completion of the optimization using SGD, the VHMR can be activated as required. Specifically, the newly received haptic signal $d_{new}^{(h)}$ is firstly processed through the haptic feature extraction and semantic association modules to obtain the corresponding representation $g_{new}^{(h)}$ in the shared semantic space.

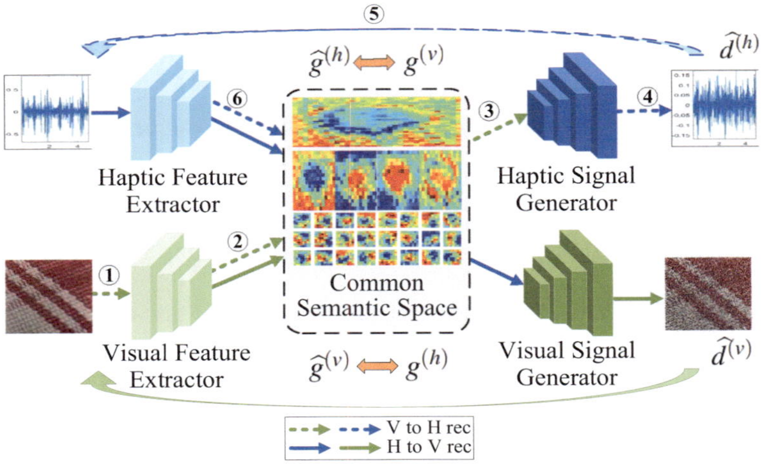

Fig. 5.11 Inter-modal cyclic information flow

Subsequently, $g_{new}^{(h)}$ is input into the trained generator GV to recover the desired visual signal $\widehat{d}_{new}^{(v)}$. Similarly, in scenarios where the received haptic signals are delayed or of low quality, the VHMR approach is automatically activated. In such cases, the relevant visual signal $d_{new}^{(v)}$ is input into the visual feature extraction and semantic association modules to obtain the representation $g_{new}^{(v)}$. Thereafter, $g_{new}^{(v)}$ is input into the trained generator GH to recover $\widetilde{d}_{new}^{(h)}$.

5.3.6 Experimental Result and Analysis

In this section, the VHMR approach is employed on a standard dataset. Its performance can be evaluated by comparing with the other competing approaches.

Experimental configuration

Dataset: It contains two large-scale visual auxiliary databases CURET [21] and ETHZ Synthesizability [22], one haptic auxiliary database constructed by haptic data with three annotated characteristics. The specific haptic characteristics are friction (rough or smooth), compliance (soft or hard), and thermal conductance (warm or cold). Moreover, it also contains LMT Dataset [6] taken as the received dataset at the edge node. 80% of the visual and haptic signal pairs from this dataset are selected for training, while the remaining 20% of samples are used for testing.

Competing schemes: To assess visual signal recovery performance of the VHMR framework, a comparative analysis is conducted with four competing schemes: Pix2Pix [23], DiscoGAN [24], CMC-GAN [25], and deep canonically correlated

Algorithm 5.3 Optimization process of the VHMR [19]

Require: Auxiliary databases $A^{(v)}$, $A^{(h)}$, received datasets $D^{(v)}$, $D^{(h)}$ for training at the edge nodes; Initial parameters $\{\theta_{kf}^{(v)}, \theta_{rc}^{(v)}, \theta_{kf}^{(h)}\}$, $\{\theta_{fe}^{(v)}, \theta_{fe}^{(h)}\}$, $\{\theta_{cm}^{(v)}, \theta_{cm}^{(h)}, \theta_{cl}^{(v)}, \theta_{cl}^{(h)}\}$, $\{\theta_{rec}^{(v)}, \theta_{dis}^{(v)}, \theta_{rec}^{(h)}, \theta_{dis}^{(h)}\}$; Batch-size, learning rate μ_1, μ_2, μ_3, iterations n_1, n_2, n_3

Stage 1: Feature extraction by knowledge transfer
for iteration $= 1, \ldots, n_1$ **do**
  Update $\theta_{kf}^{(v)}, \theta_{rc}^{(v)}$: $\theta_{kf}^{(v)} \leftarrow \theta_{kf}^{(v)} - \mu_1 \nabla_{\theta_{kf}^{(v)}} \mathcal{L}_{AV}$, $\theta_{rc}^{(v)} \leftarrow \theta_{rc}^{(v)} - \mu_1 \nabla_{\theta_{rc}^{(v)}} \mathcal{L}_{AV}$
  Update $\theta_{kf}^{(h)}$: $\theta_{kf}^{(h)} \leftarrow \theta_{kf}^{(h)} - \mu_1 \nabla_{\theta_{kf}^{(h)}} \mathcal{L}_{AH}$, $\theta_{rc}^{(h)} \leftarrow \theta_{rc}^{(h)} - \mu_1 \nabla_{\theta_{rc}^{(h)}} \mathcal{L}_{AH}$
  Update $\theta_{fe}^{(v)}, \theta_{fe}^{(h)}$: $\theta_{fe}^{(v)} \leftarrow \theta_{fe}^{(v)} - \mu_1 \nabla_{\theta_{fe}^{(v)}} \mathcal{L}_{tran}$, $\theta_{fe}^{(h)} \leftarrow \theta_{fe}^{(h)} - \mu_1 \nabla_{\theta_{fe}^{(h)}} \mathcal{L}_{tran}$
end for
return $k^{(v)}, k^{(h)}, f^{(v)}, f^{(h)}$
Stage 2: Semantic association & Coarse-grained signal recovery
for iteration $= 1, \ldots, n_2$ **do**
  Update $\theta_{cm}^{(v)}, \theta_{cm}^{(h)}, \theta_{cl}^{(v)}, \theta_{cl}^{(h)}$: $\theta_{cm}^{(v)} \leftarrow \theta_{cm}^{(v)} - \mu_2 \nabla_{\theta_{cm}^{(v)}} (\mathcal{L}_{cor} + \alpha \mathcal{L}_{cat})$, $\theta_{cm}^{(h)} \leftarrow \theta_{cm}^{(h)} - \mu_2 \nabla_{\theta_{cm}^{(h)}} (\mathcal{L}_{cor} + \alpha \mathcal{L}_{cat})$, $\theta_{cl}^{(v)} \leftarrow \theta_{cl}^{(v)} - \mu_2 \nabla_{\theta_{cl}^{(v)}} (\mathcal{L}_{cor} + \alpha \mathcal{L}_{cat})$, $\theta_{cl}^{(h)} \leftarrow \theta_{cl}^{(h)} - \mu_2 \nabla_{\theta_{cl}^{(h)}} (\mathcal{L}_{cor} + \alpha \mathcal{L}_{cat})$
  Update $\theta_{rec}^{(v)}, \theta_{dis}^{(v)}$: $\theta_{rec}^{(v)} \leftarrow \theta_{\mathrm{rec}}^{(v)} - \mu_2 \nabla_{\theta_{rec}^{(v)}} (\mathcal{L}_{GV} + \beta_V \mathcal{L}_{PV})$, $\theta_{dis}^{(v)} \leftarrow \theta_{dis}^{(v)} - \mu_2 \nabla_{\theta_{dis}^{(v)}} \mathcal{L}_{DV}$
  Update $\theta_{rec}^{(h)}, \theta_{dis}^{(h)}$: $\theta_{rec}^{(h)} \leftarrow \theta_{rec}^{(h)} - \mu_2 \nabla_{\theta_{rec}^{(h)}} (\mathcal{L}_{GH} + \beta_H \mathcal{L}_{PH})$, $\theta_{dis}^{(h)} \leftarrow \theta_{dis}^{(h)} - \mu_2 \nabla_{\theta_{dis}^{(h)}} \mathcal{L}_{DH}$
end for
return $g^{(v)}, g^{(h)}, \widetilde{d}^{(v)}, \widetilde{d}^{(h)}$
Stage 3: Fine-grained signal recovery
for iteration $= 1, \ldots, n_3$ **do**
  Update $\theta_{\mathrm{rec}}^{(v)}, \theta_{\mathrm{rec}}^{(h)}$: $\theta_{rec}^{(v)} \leftarrow \theta_{rec}^{(v)} - \mu_3 \nabla_{\theta_{rec}^{(v)}} \left(\lambda_V \mathcal{L}_{cycV} + \lambda_H \mathcal{L}_{cycH}\right)$, $\theta_{rec}^{(h)} \leftarrow \theta_{rec}^{(h)} - \mu_3 \nabla_{\theta_{rec}^{(h)}} \left(\lambda_V \mathcal{L}_{cycV} + \lambda_H \mathcal{L}_{cycH}\right)$
end for
return $\widehat{d}^{(v)}, \widehat{d}^{(h)}$

autoencoder (DCCAE) [26]. In this evaluation, the input to each scheme consists of haptic spectrograms derived from the corresponding haptic signals. Additionally, the FHIS proposed in Sect. 5.2 is included as a benchmark for comparison.

To evaluate the haptic signal recovery efficacy of the VHMR, a comparison is made with the following schemes: Pix2Pix, simple regression (SR), DiscoGAN, and deep visual-haptic learning (DVHL) [27]. The haptic spectrograms generated by these schemes were subsequently converted into final haptic signals using the Griffin-Lim algorithm [28]. Furthermore, the AVHR proposed in Sect. 5.1 is selected as an additional competing scheme for this evaluation.

Implementation details: The implementation details of all the modules and networks in the VHMR are listed in the following.

Visual knowledge acquisition network: It is constructed based on four 4×4 convolutional layers with channel dimensions of 32, 64, 128, and 256, and then followed by a two-layer fully connected neural network with dimensions of 128 and 64.

Haptic knowledge acquisition network: A stacked autoencoder with dimensions of 1024, 512, and 256 is employed, and then succeeded by a two-layer fully connected neural network with dimensions of 128 and 64. Finally, a fully connected layer with a softmax activation function is utilized to construct the classifier.

Visual & haptic feature extraction and semantic association modules: The dimensionality of the output features is 64. Furthermore, two feedforward neural networks with dimensions of 128 and 64 are selected to map the features into the shared semantic space. Moreover, two fully connected layers with softmax activation functions are employed to perform classification tasks.

Visual & haptic signal recovery modules: For the visual recovery module, the generator consists of a two-layer fully connected network followed by a four-layer deconvolutional network with channel dimensions of 256, 128, 64, and 32, respectively. The discriminator is a three-layer fully connected network with dimensions of 512, 256, and 1. For the haptic recovery module, the generator is the decoder part of the stacked autoencoder with dimensions of 64, 128, 256, 512, and 1024. The discriminator is a three-layer fully connected network with dimensions of 512, 256, and 1.

Evaluation metrics: To evaluate the performance of visual signal recovery, three metrics are employed: ACC, IS, and structural similarity index measure (SSIM). Specifically, ACC is assessed by pre-training a classifier on a real visual dataset. Subsequently, the recovered visual signal is fed into this pre-trained classifier to determine if its category matches that of the real visual signal. The inception score (IS) is utilized to quantitatively evaluate the quality and diversity of the recovered visual signals $\widehat{d}^{(v)}$ [10]. SSIM measures the structural similarity of visual signals by considering three aspects: luminance, contrast, and structure. The SSIM value ranges from 0 to 1, with values closer to 1 indicating less distortion in the visual signal [29].

For verifying the performance of haptic signal recovery, ACC, RMSE, and spectral-temporal similarity index measure (ST-SIM) are utilized. The ACC metric is identical to the one described above. RMSE is employed to evaluate the one-to-one correspondence between $\widehat{d}_i^{(h)}$ and $d_i^{(h)}$. ST-SIM measures the quality of the recovered haptic signals by assessing both perceptual spectral and temporal similarity [30].

Results and analysis

Visual signal recovery result: Table 5.6 presents the performance metrics for visual signal recovery. With respect to ACC, the VHMR approach exhibits a markedly superior discriminative capacity when compared with four alternative methods. Furthermore, when appraised through the lenses of IS and SSIM, the VHMR approach demonstrates robust performance in both the quality and diversity of the recovered visual signals. This result of the VHMR can be attributed to three salient factors. Primarily, the knowledge transfer from a large-scale auxiliary visual database significantly enhances the efficacy of visual feature extraction. Secondly, the VHMR approach adeptly explores semantic consistency across different modalities. Lastly, the optimization of the generator through cyclic-consistency constraints markedly improves the quality of the recovered visual signals. When compared with the FHIS approach in Sect. 5.2, the performance of the VHMR is marginally inferior. This discrepancy arises because the FHIS approach is specifically tailored for a particular

Table 5.6 Visual recovery performance comparison of the VHMR and the other competing approaches

Approach	ACC	IS	SSIM
Original	–	10.096 ± 0.147	–
Pix2Pix	0.594	4.917 ± 0.357	0.433
DiscoGAN	0.667	6.012 ± 0.113	0.595
CMC-GAN	0.640	5.159 ± 0.218	0.556
DCCAE	0.497	5.176 ± 0.196	0.423
FHIS	0.763	9.432 ± 0.976	0.798
VHMR	**0.749**	$\mathbf{7.154 \pm 0.195}$	**0.765**

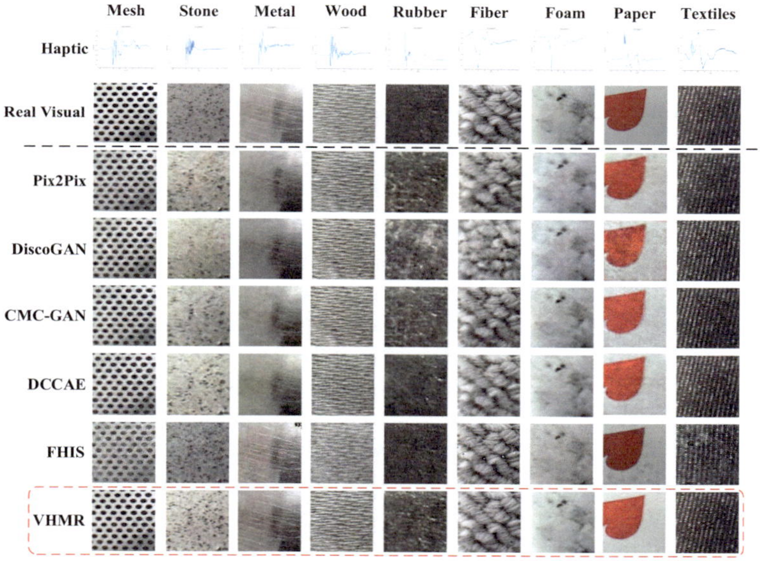

Fig. 5.12 Recovery effects of the visual signals by the VHMR and the other competing approaches

haptic-aided visual recovery task and does not address the challenge of visual-haptic mutual recovery, which is a distinct advantage of the proposed VHMR approach.

In addition to quantitative evaluation, a qualitative comparison of the recovery effects was conducted using the VHMR approach across nine material categories. Figure 5.12 illustrates the received haptic signals, real visual signals, and recovered visual signals obtained using the VHMR approach and other competing methods. It can be seen that the recovered visual signals by the other approaches have differences with real ones (e.g. "stone", "wood", "rubber", "foam"). In contrast, the VHMR approach can obtain high-quality visual signals, albeit with a discernible gap when compared to the real visual signals.

Table 5.7 Haptic recovery performance comparison of the VHMR and the other competing approaches

Approach	ACC	RMSE	ST-SIM
SR	0.378	2.466	0.587
Pix2Pix	0.515	3.597	0.658
DiscoGAN	0.598	2.049	0.762
DVHL	0.498	3.601	0.684
AVHR	0.716	0.883	0.842
VHMR	**0.684**	**0.872**	**0.836**

Haptic signal recovery result: Table 5.7 delineates the performance metrics associated with haptic signal recovery. Upon examination of ACC and RMSE metrics, it is evident that the haptic signals recovered via the VHMR approach exhibit markedly superior discrimination and fidelity. Moreover, ST-SIM reveals that the haptic signals generated by the VHMR possess enhanced quality in terms of both perceptual spectral and temporal similarity. These outcomes can be ascribed to three principal factors. Initially, the knowledge transfer from the haptic database facilitates more effective haptic feature extraction. Secondly, the incorporation of semantic information derived from visual signals further augments efficacy of the recovered haptic signals. Lastly, the optimization process extends beyond mere focus on haptic spectrograms to encompass the recovered haptic signals themselves, thereby improving overall recovery quality. When compared with the AVHR presented in Sect. 5.1, the performance of the VHMR is marginally inferior. This disparity is attributable to the fact that the AVHR is specifically designed for a particular haptic-aided visual recovery task. In other words, the AVHR does not address the broader challenge of visual-haptic mutual recovery, which is a fundamental requirement for equally important multi-modal services.

In addition to quantitative analysis, a qualitative comparison of haptic recovery effects was conducted using the VHMR approach across nine material categories. Figure 5.13 presents the received visual signals, real haptic signals, and recovered haptic signals obtained using VHMR and other competing methods. It is evident that the recovered haptic signals by the VHMR closely approximate the real ones, (e.g. “metal”, “wood”, “rubber”).

Ablation experimental analysis: To further substantiate the efficacy of the core modules within the VHMR approach, an ablation study has been meticulously designed. This study involves the formation of several baseline configurations by selectively removing different modules. The performance of these configurations is evaluated using SSIM metric for the recovered visual signals and ST-SIM metric for the recovered haptic signals. The results are presented in Table 5.8. The configuration labeled “VHMR w/o KF” represents a scenario where knowledge acquisition and transfer mechanisms are disabled. By comparing the performance of “VHMR w/o KF” with that of the fully enabled “VHMR (fully)”, it is evident that the knowl-

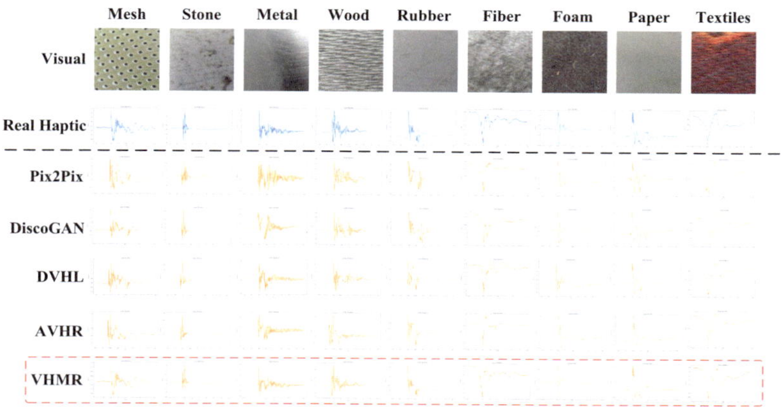

Fig. 5.13 Recovery effects of the haptic signals by the VHMR and the other competing approaches

Table 5.8 Result about impacts of each module in the VHMR

Approach	SSIM (visual)	ST-SIM (haptic)
VHMR w/o KF	0.683	0.744
VHMR w/o SA	0.614	0.693
VHMR w/o FR	0.709	0.798
VHMR (fully)	**0.765**	**0.836**

edge transferred from auxiliary visual and haptic databases significantly enhances feature extraction, thereby effectively improving the overall recovery performance. The configuration "VHMR w/o SA" indicates a scenario where semantic association is not considered. As shown in Table 5.8, both the SSIM for visual signals and the ST-SIM for haptic signals exhibit inferior performance in this configuration. This observation underscores the critical role of semantic association in the subsequent coarse-grained and fine-grained signal recovery processes. The configuration "VHMR w/o FR" denotes a scenario where fine-grained signal recovery is omitted. The results reveal a noticeable degradation in performance compared to the fully enabled "VHMR (fully)". This suggests that by leveraging one modality signal to assist in the recovery of the other and imposing a semantic consistency constraint on the generated feedback signals, the quality of the recovered visual and haptic signals can be significantly enhanced.

Performance under different channel environments: To assess the performance of the VHMR approach under various challenging channel conditions, including noise, fading, and interference, a comprehensive experiment has been conducted. Here, neural networks are employed to approximate the intricate characteristics of additive white Gaussian noise (AWGN) and Rayleigh channels. This choice is motivated by the neural networks' capability to effectively model the complex nonlinear mapping relationships between the input and output of these specific channels [31].

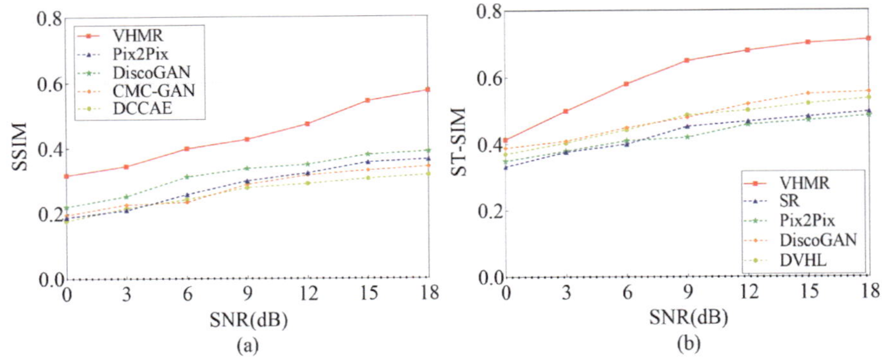

Fig. 5.14 Recovery effect comparison under AWGN channel: **a** Quality of visual signals by SSIM; **b** Quality of haptic signals by ST-SIM

The recovery quality of visual and haptic signals under an AWGN channel is illustrated in Fig. 5.14. As depicted in Fig. 5.14a, with the increase in SNR, indicating improved channel conditions, the quality of the generated visual signals correspondingly enhances. Notably, SSIM of the VHMR approach exceeds 0.6, whereas SSIM values of other competing methods remain below 0.5. This demonstrates the superior performance of VHMR in visual signal recovery under AWGN conditions. Figure 5.14b reveals a significant performance gap between the VHMR and other competing approaches in terms of haptic signal recovery under the AWGN channel. The VHMR consistently outperforms its counterparts, highlighting its robustness and effectiveness in maintaining high-quality haptic signal recovery even in the presence of noise.

The recovery quality of visual and haptic signals under a Rayleigh channel is presented in Fig. 5.15. It is important to note that the transmitted multi-modal streams are subject to multi-path propagation and multiple random fading factors in a Rayleigh channel environment. Consequently, the recovery performance under a Rayleigh channel is generally inferior to that under an AWGN channel. Despite these challenges, the proposed VHMR approach exhibits superior visual and haptic recovery performance compared to other competing methods. This underscores the resilience and adaptability of VHMR in handling complex channel conditions, thereby ensuring reliable multi-modal signal recovery.

5.4 Information Retrieval-Based Signal Recovery

Essentially, the above three sections utilize the received undamaged signals to recover the damaged ones. Unlike them, information retrieval is another important approach that utilizes the successfully received or already cached data to directly compensate for the loss, delayed arrival of certain modality data packets, or data asynchronization

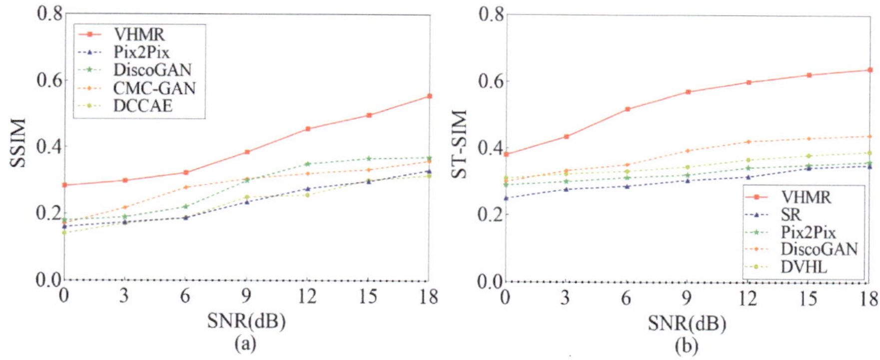

Fig. 5.15 Recovery effect comparison under Rayleigh channel: **a** Quality of visual signals by SSIM; **b** Quality of haptic signals by ST-SIM

at the receiver. However, there does not exist the works about information retrieval-based signal recovery in communication environments. The main reasons are three key challenges: Firstly, it needs to construct a framework for signal recovery by fully leveraging these successfully received or already cached data. Secondly, solving the mutual retrieval problem among visual, audio, and haptic modality signals is extremely difficult. Thirdly, the recovery performance in the different wireless channels needs to be carefully checked. To address the above challenges, this section proposes an information retrieval-based signal recovery (IRSR) scheme for cross-modal communications [32]. Specifically, a framework is firstly proposed. At the sender, visual, audio and haptic signals are preprocessed and delivered. Considering the channel noise pollution during transmission, at the receiver, the cached or successfully received multi-modal data can be used to achieve information recovery. Then, a signal recovery method based on mutual retrieval among modalities is designed, containing intra-modal one-to-one retrieval, inter-modal one-to-one retrieval, and inter-modal one-to-many retrieval. Finally, recovery performances under different SNRs and bandwidth limitations are evaluated.

5.4.1 Overall Framework

The overall framework of the IRSR approach is shown in Fig. 5.16. Firstly, the preprocessed visual, audio, and haptic signals are input into a VGG16 network, from which the fully connected layer at the end is removed, and then flattened to obtain one-dimensional vectors as the features of each modality. Subsequently, the features output from the flattening layer of the three modalities are trained separately to learn better features. Next, all the modalities are jointly input into the semantic fusion module for training to achieve the interconnection of deep semantics across different modalities. Finally, retrieval is performed to realize information recovery.

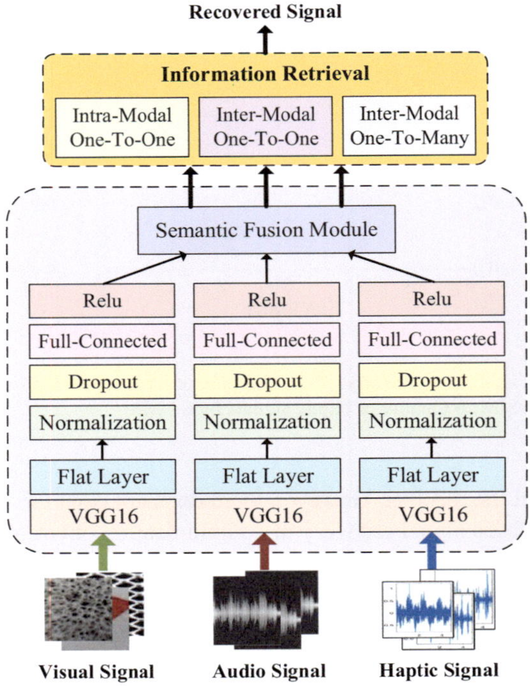

Fig. 5.16 Framework of the IRSR

When concerning the retrieval, different methods are employed based on the specific conditions during the multi-modal transmission. Here, it is assumed that the conditions of the transmission process and the corresponding retrieval methods are known. Specifically, the visual, audio, and haptic signals are denoted as V, A, and H, respectively, and the set of all modalities is denoted as $All = \{V, A, H\}$. Three typical scenarios are considered: intra-modal one-to-one retrieval, including $V \to V$, $A \to A$, and $H \to H$; inter-modal one-to-one retrieval, including $V \to A$, $V \to H$, $A \to V$, $A \to H$, $H \to V$, $H \to A$; inter-modal one-to-many retrieval, including $V \to All$, $A \to All$, $H \to All$.

5.4.2 *Multi-modal Feature Extraction*

Prior to semantic fusion of different modalities, it is necessary to train each modality separately to further extract features. The network structure for feature extraction contains a batch normalization layer (BN), dropout layer 1, fully connected layer 1 (FC), the activation function ReLU, then following dropout layer 2, fully connected layer 2, and the softmax function. Specifically, the BN layer can normalize the data, accelerate model training, and has a regularization effect. The dropout layers can prevent model overfitting. The fully connected layers are used to learn features. The activation function ReLU can enhance the network's nonlinear mapping capability, and the softmax function is used for data classification. After the network is trained, dropout layer 2, fully connected layer 2, and the softmax function are removed. The remaining structure is connected to the semantic fusion module.

The feature extraction for visual, audio, and haptic modalities all employ the multi-class cross-entropy loss function for their respective network training. Take the loss function of visual modality is f_V as an example,

$$f_V = \frac{1}{N_V} \sum_{k=1}^{N_V} \sum_{j=1}^{C_V} 1\left\{y_V^k = j\right\} \log\left[p\left(x_V^k\right)\right], \tag{5.29}$$

where N_V denotes the total number of samples in the visual modality training data. x_V^k represents the feature extracted from the k-th sample of the visual modality after passing through the softmax function in the feature extraction structure. y_V^k indicates the label of the k-th sample of the visual modality, with the label values being $1, 2, \ldots, C_V$. C_V also signifies the total number of categories in the visual modality. $1\{y_V^k = j\}$ is an indicator function, which equals 1 when y_V^k belongs to label j, and 0 otherwise. $p(x_V^k)$ represents the probability that the label of x_V^k is j. The expressions of f_A and f_H are similar to that of f_V.

5.4.3 Semantic Fusion Among Different Modality Features

After feature extraction, features of three modalities are jointly fed into semantic fusion network. More importantly, a loss function has been designed for this module to conduct semantic correlation among different modalities. The loss function consists of two parts, $L_{cls} + \lambda \cdot L_{cen}$. Here, the first item L_{cls} is about the specific-modal loss, that is

$$L_{cls} = \frac{1}{N_V} \sum_{k=1}^{N_V} g\left(x_V^k, y_V^k\right) + \frac{1}{N_A} \sum_{k=1}^{N_A} g\left(x_A^k, y_A^k\right) + \frac{1}{N_H} \sum_{k=1}^{N_H} g\left(x_H^k, y_H^k\right). \tag{5.30}$$

The formations of $g(x_V^k, y_V^k)$, $g(x_A^k, y_A^k)$, and $g(x_H^k, y_H^k)$ can be described by the multi-class cross-entropy loss function. The second item is about the cross-modal loss,

$$L_{cen} = \frac{1}{2} \sum_{m=1}^{N} \left\| x^m - c_m \right\|_2^2. \tag{5.31}$$

Here, N represents the total number of training data samples about visual, audio, and haptic modalities, that is, $N = N_v + N_A + N_H$. x^m denotes the feature of the

m-th training data, and c_m is the center of the corresponding category of the m-th training data, which dynamically changes during model training process. By utilizing this loss, it can make the same categories of different modalities more compact. By optimizing $L_{cls} + \lambda \cdot L_{cen}$, the parameters of the semantic fusion network can be determined.

5.4.4 Cross-Modal Information Retrieval

After employing the above procedures, a query sample q_i and a retrieval sample r_j are defined according to the three cases.

Case 1: In this case, the visual, audio, and haptic signals are all corrupted by channel noise within a certain period. Taking the visual modality as an example, the noise-corrupted visual signal is used as the query sample to retrieve a similar visual signal from the database near the receiver. The retrieved similar visual signal is then used to replace the noise-corrupted one, achieving the purpose of signal recovery. This is known as intra-modal one-to-one retrieval. Therefore, $q_i \in V \cup r_j \in V, q_i \in A \cup r_j \in A, q_i \in H \cup r_j \in H$.

Case 2: In this case, one modality signal is lost within a certain period. Assuming the visual signal is lost, the audio is used as the query sample to retrieve a visual signal from the database near the receiver that is most similar in category to the query sample. This most similar visual signal is then used to substitute for the lost one, achieving the purpose of signal recovery. This is known as inter-modal one-to-one retrieval. Therefore, $q_i \in V \cup r_j \in A$, $q_i \in V \cup r_j \in H$, $q_i \in A \cup r_j \in V$, $q_i \in A \cup r_j \in H$, $q_i \in H \cup r_j \in V$, $q_i \in H \cup r_j \in A$.

Case 3: In this case, two modality signals may be lost within a certain period. For example, the visual and audio signals might be lost. However, the haptic signal is still available. In this situation, the haptic signal is used as the query sample. The system then retrieves visual and audio signals from the database near the receiver. These retrieved signals are similar in category to the query sample. These retrieved signals are then used to substitute for the lost visual and audio signals. This is known as inter-modal one-to-many retrieval. Therefore, $q_i \in V \cup r_j \in All, q_i \in A \cup r_j \in All$, $q_i \in H \cup r_j \in All$.

The cosine similarity function is used to measure the distance between two vectors, as shown in Eq. (5.32). Here, i and j represent the indices of the query sample and the retrieval sample, respectively. By fixing i and iterating j through the received or caching dataset, the samples are then sorted in descending order of distance. A larger distance indicates greater similarity between the two samples. The most similar results are outputted, replacing the signal that is noisy or missing, thus achieving signal recovery

$$Sim\langle q_i, r_j\rangle = \frac{q_i \cdot r_j}{\|q_i\| \cdot \|r_j\|}. \tag{5.32}$$

5.4.5 Experimental Result and Analysis

In order to evaluate the performance of this approach, the LMT database is selected, while the samples of each material are shuffled and then redivided into training, validation, and testing sets in the ratio of 3:1:1. The testing set is used for the final retrieval and the effectiveness of signal recovery is verified based on the metric mean average precision (MAP). For each query sample, the average precision is calculated based on all samples in the retrieval set, and then the mean of the AP values for all query samples is taken to obtain the final MAP value,

$$MAP = \frac{1}{Q}\sum_{j=1}^{Q}\frac{\sum_{i=1}^{R} Pre(i)Rel(i,j)}{\sum_{i=1}^{R} Rel(i,j)}, \tag{5.33}$$

where Q is the total number of query samples. R is the total number of samples in the retrieval set. $Pre(i)$ is the precision of the sample retrieved at position i, and $Rel(i, j)$ is the relevance of the query sample j to the retrieved sample at position i. If they belong to the same category, then $Rel(i, j) = 1$, otherwise, $Rel(i, j) = 0$.

When concerning the implementation details, the preprocessed three modality signals can be unified as "image" and are adjusted to the same resolution. The size of the wireless channel bandwidth B is indirectly reflected by the difference in resolution. This experiment considers the resolutions $128 \times 128 \times 3$, $128 \times 96 \times 3$, and $128 \times 64 \times 3$ as scenarios with bandwidths of 128, 96, and 64, respectively. The SNR range discussed is from –12 to 15 dB, as well as the ideal channel scenario ($B = 128$, $SNR = \infty$).

Firstly, the IRSR approach is compared with four traditional machine learning methods and three deep learning methods. These methods include canonical correlation analysis (CCA), kernel canonical correlation analysis (KCCA), principal component analysis (PCA), independent component analysis (ICA), autoencoder (AE), variational autoencoder (VAE), and attention mechanism. For KCCA, the Gaussian kernel function is used as the kernel. The experimental results shown in Table 5.9 indicate that the MAP value of the proposed method is significantly higher than those of the other competing methods.

Subsequently, an ablation experiment is conducted. A comparison is made among three methods: the proposed approach, the proposed approach without feature extraction module (denoted as "IRSR w/o FE"), the proposed approach without semantic fusion (denoted as "IRSR w/o SF"). Table 5.10 presents the result. It can be observed that, without individual network training for extracting features of different modalities, the retrieval performance deteriorates significantly. Similarly, in the absence of semantic fusion module, that is, when the retrieval is conducted based on the uncorrelated features of the three modalities, the retrieval performance is also far inferior to that of the proposed approach.

Finally, the impact of bit errors during transmission on experimental performance is considered. Specifically, it is assumed that the training and testing sets have the

Table 5.9 The MAP comparison of the proposed and the other competing approaches [32]

Approach	$V \to V$	$V \to A$	$V \to H$	$A \to V$	$A \to A$	$A \to H$
CCA	0.179	0.140	0.138	0.143	0.168	0.139
KCCA	0.199	0.198	0.199	0.201	0.217	0.205
PCA	0.302	0.144	0.152	0.161	0.250	0.159
ICA	0.297	0.147	0.144	0.152	0.247	0.178
AE	0.303	0.154	0.165	0.154	0.242	0.151
VAE	0.296	0.149	0.155	0.151	0.248	0.161
Attention	0.673	0.453	0.367	0.485	0.425	0.338
IRSR	**0.921**	**0.702**	**0.588**	**0.712**	**0.560**	**0.474**
Approach	$H \to V$	$H \to A$	$H \to H$	$V \to All$	$A \to All$	$H \to All$
CCA	0.139	0.138	0.185	0.141	0.140	0.146
KCCA	0.200	0.216	0.216	0.201	0.205	0.205
PCA	0.172	0.151	0.279	0.183	0.165	0.175
ICA	0.140	0.178	0.268	0.188	0.175	0.181
AE	0.157	0.145	0.290	0.188	0.162	0.178
VAE	0.153	0.154	0.295	0.187	0.170	0.186
Attention	0.420	0.353	0.338	0.506	0.403	0.355
IRSR	**0.636**	**0.479**	**0.451**	**0.762**	**0.580**	**0.510**

Table 5.10 The MAP comparison of the proposed and the degraded approaches

Approach	$V \to V$	$V \to A$	$V \to H$	$A \to V$	$A \to A$	$A \to H$
IRSR w/o FE	0.217	0.164	0.167	0.164	0.164	0.164
IRSR w/o SF	0.477	0.127	0.149	0.143	0.363	0.149
IRSR (fully)	**0.921**	**0.702**	**0.588**	**0.712**	**0.560**	**0.474**
Approach	$H \to V$	$H \to A$	$H \to H$	$V \to All$	$A \to All$	$H \to All$
IRSR w/o FE	0.167	0.164	0.164	0.157	0.136	0.136
IRSR w/o SF	0.163	0.148	0.355	0.259	0.217	0.219
IRSR (fully)	**0.636**	**0.479**	**0.451**	**0.762**	**0.580**	**0.510**

same SNR. Figures 5.17 and 5.18 illustrate the relationship between bandwidth B, SNR, and MAP value under Rayleigh channel conditions. On one hand, as the SNR of the testing samples increases, MAP value also increases, indicating the retrieval performance improvement. On the other hand, when SNR of the training samples does not vary significantly, MAP value remains relatively stable, suggesting similar retrieval-based signal recovery performance. However, when SNR of the training samples is relatively low, for example, at –3 dB, the curve shows a severe degradation

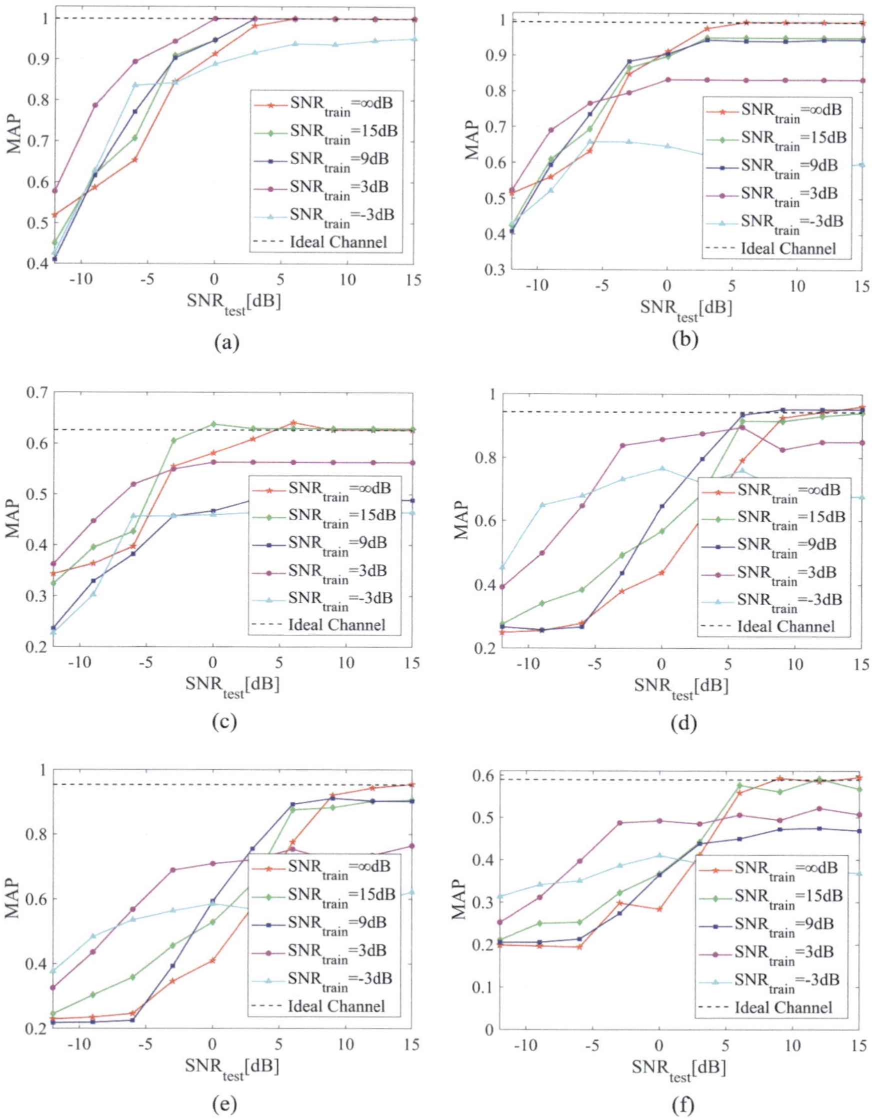

Fig. 5.17 MAP values under different Rayleigh channels: **a** $V \rightarrow V$; **b** $V \rightarrow A$; **c** $V \rightarrow H$; **d** $A \rightarrow V$ **e** $A \rightarrow A$; **f** $A \rightarrow H$

as SNR of the testing samples increases. Therefore, if the proposed approach cannot be trained with a sufficient amount of high-quality data, it is possible to moderately reduce the quality of the training data, which can still yield satisfactory retrieval-based signal recovery performance.

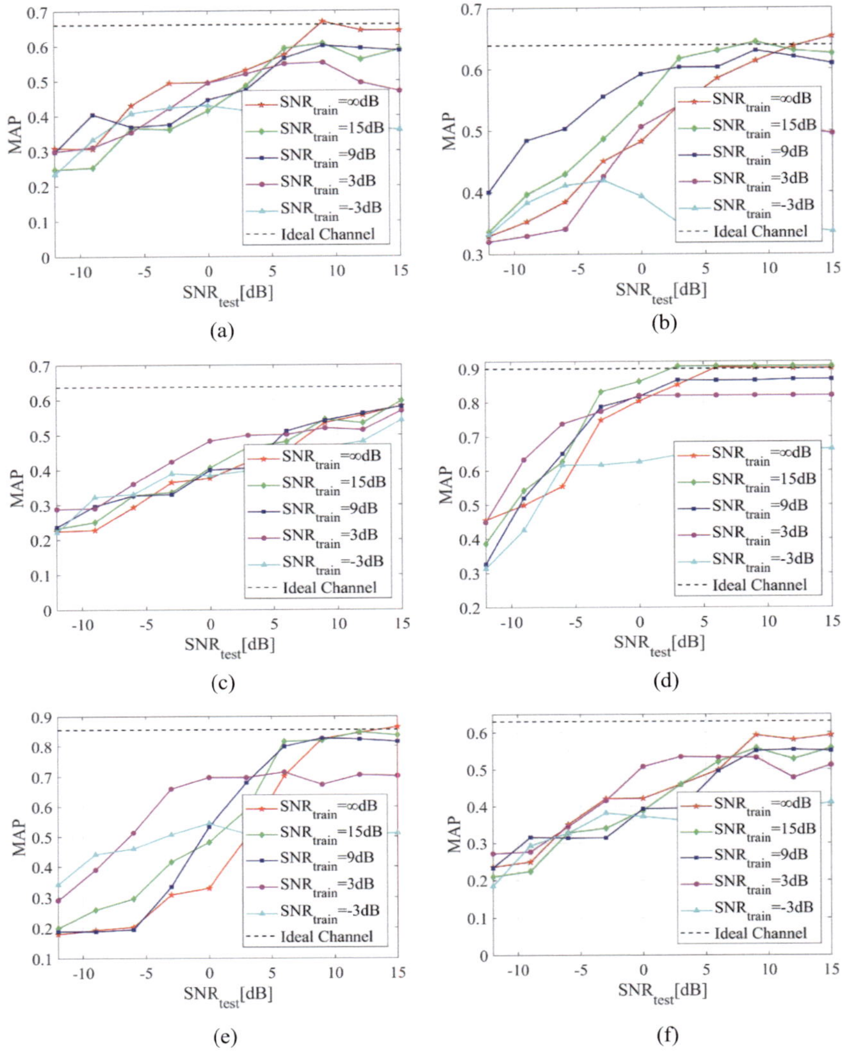

Fig. 5.18 MAP values under different Rayleigh channels: **a** $H \rightarrow V$; **b** $H \rightarrow A$; **c** $H \rightarrow H$; **d** $V \rightarrow All$ **e** $A \rightarrow All$; **f** $H \rightarrow All$

5.5 Cross-Modal Super-Resolution Signal Reconstruction

In certain multi-modal services such as remote industrial robotic grasping, the constraints of bandwidth and the presence of inter-modal competition pose significant challenges for effective cross-modal communications. Firstly, the extensive deployment of devices at slave domain results in high concurrency. This makes the transmission of large volumes of high-resolution (HR) images and videos over limited bandwidth extremely difficult. It thereby degrades the immersive experience for users, particularly in terms of visual quality, and potentially disrupts the smooth functioning of machinery. Secondly, when the high-priority haptic signals emerge, the transmission efficiency of large-scale visual signals is adversely affected. To tackle these challenges, a potential solution is to transmit only low-resolution (LR) visual signals and reconstruct the HR ones at the receiver. However, existing methods typically perform super-resolution reconstruction solely at local terminals [33]. For remote industrial robotic grasping service, multi-modal signals must be transmitted over wireless networks, which are susceptible to issues such as noise and interference. Against this backdrop, a super-resolution reconstruction strategy for cross-modal communications is proposed [34]. It involves transmitting only LR visual signals and haptic signals from the sender and reconstructing the corresponding HR visual signals at the receiver. To be more specific, a haptic-aided super-resolution reconstruction (HaSR) approach is developed. This approach encompasses the extraction of features from visual and haptic signals, semantic learning and fusion based on both intra-modal and inter-modal constraints, and the reconstruction of HR images. Experimental results demonstrate the efficacy of the proposed approach.

5.5.1 Overall Framework

Here, we aim to generate HR visual signals by leveraging haptic signals and LR visual signals. Consequently, during the training phase, HR images, their corresponding LR images, and haptic signals are required. Then, in the subsequent reconstruction phase, the trained model can utilize the given haptic signals and LR images from the slave domain to generate HR images at the master domain. It is important to note that in remote industrial robotic grasping, haptic signals and images generally have a one-to-one mapping, as images can be captured simultaneously while touching the object. Additionally, timestamps for synchronization are set in the relevant packet headers before transmitting the visual and haptic streams, ensuring the one-to-one mapping of LR visual and haptic signals. Assuming that the training database $D = \{d_i\}_{i=1}^{N}$ contains N instances, each instance d_i consists of an HR image h_i, a low-resolution image l_i, and a haptic signal t_i. Framework of the HaSR is shown in Fig. 5.19. It is primarily composed of two parts. The first part is analogous to an autoencoder, where features are extracted from the HR image using a HR image encoder E_h, and then fed into a decoder that consists of a HR image generator G_h and a discriminator

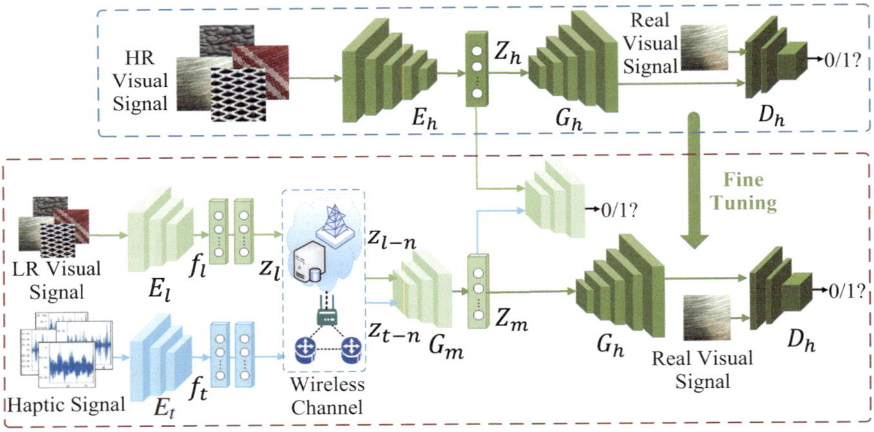

Fig. 5.19 Framework of the HaSR

D_h. The encoder is a convolutional neural network specifically designed to learn and extract high-level semantic information, which serves as both a target and a constraint for processing LR images and haptic signals. The second part comprises LR image and haptic signal encoding networks E_l and E_t, mapping networks F_l and F_t, and a feature fusion network F_m, which includes a fused feature generator G_m and a discriminator D_m.

5.5.2 *High Resolution (HR) Visual Encoding and Decoding Before Delivery*

To reconstruct HR images using haptic signals and LR images, we initially employ an autoencoder to perform the encoding and decoding of HR images. This approach allows us to obtain the potential features within HR images along with the corresponding decoding network. These components serve as a feature constraint and provide a pre-trained decoder for the subsequent stages. Given the exceptional capability of CNN in image feature extraction, they are selected as the encoder. Drawing on the recent successful application of GAN in image generation [36], we utilize it as the decoder. The loss function can be expressed as

$$L_{pre} = L_{rec} + \alpha L_{pre-adv}, \tag{5.34}$$

where α is a coefficient that adjusts the proportion of different loss terms. The first loss L_{rec} represents the reconstruction loss.

$$L_{rec} = \frac{1}{C} \left\| G_h \left(z_h^i \right) - h_i \right\|_1, \tag{5.35}$$

where the product of C indicates the size of the HR images. z_h^i denotes the potential features of the ith HR image, and $\| \cdot \|_1$ signifies the corresponding L_1 norm. The second loss $L_{pre-adv}$ is the adversarial loss. We opt for Wasserstein GAN (WGAN) [37] to enhance the stability of network training. $L_{pre-adv}$ can be formulated as

$$\min_{\theta_{gh}} \max_{\theta_{dh}} L_{pre-adv} (G_h, D_h) = -E_{z_h \sim p(z_h)} \left(D_h \left(G_h \left(z_h \right) \right) \right) + E_{h \sim p(h)} \left(D_h (h) \right), \tag{5.36}$$

where z_h represents the coding features of the HR images. θ_{gh} and θ_{dh} respectively correspond to the parameters of the generator and discriminator.

5.5.3 *Feature Extraction and Mapping for Low Resolution (LR) Visual and Haptic Signals*

Leveraging the coding features of the HR images z_h acquired in the preceding step, the objective is to utilize the available LR images and haptic signals to approximate z_h. To this end, it is imperative to employ effective strategies to bridge the semantic gap and investigate semantic correlation between visual and haptic modalities. This can be accomplished through a two-step process. Initially, preliminary feature extraction is conducted via the encoding networks dedicated to haptic signals and low-resolution images, respectively yielding the haptic encoding feature f_t and the visual coding feature f_l. Subsequently, to effectively mitigate the heterogeneity between different modalities, a feature mapping network is constructed. The overarching goal of this network is twofold: to learn the correlation between paired samples originating from distinct modalities and to discern between samples of different classes within the same modality. This dual objective is primarily realized through the optimization of the following two goals.

Inter-modal semantic relevance learning: For f_t and f_l, they are then transformed through feature mapping networks, represented as $z_l = F_l(f_l, \theta_l)$ and $z_t = F_t(f_t, \theta_t)$. θ_l and θ_t denote the parameters of the LR image mapping network $F_l(\cdot)$ and $F_t(\cdot)$, respectively. Here, the correlation between data from the same category or class but different modalities should surpass the correlation between data from different modalities and different categories. Adhering to this principle, the triplet losses to regulate the feature mapping networks of LR images and haptic signals are employed:

$$L_{cor}^{l}\left(\theta_{l}\right)=\sum_{p\neq q}^{N,N}\max\left(0,L_{2}\left(z_{l}^{p},z_{t}^{p}\right)+\sigma-L_{2}\left(z_{l}^{p},z_{t}^{q}\right)\right),$$
$$L_{cor}^{t}\left(\theta_{t}\right)=\sum_{p\neq q}^{N,N}\max\left(0,L_{2}\left(z_{t}^{p},z_{l}^{p}\right)+\sigma-L_{2}\left(z_{t}^{p},z_{l}^{q}\right)\right), \tag{5.37}$$

where p and q represent different categories. N denotes the number of data and σ is the threshold. Therefore, the loss function of inter-modal semantic relevance learning can be represented as

$$L_{cor}\left(\theta_{l},\theta_{t}\right)=L_{cor}^{l}\left(\theta_{l}\right)+L_{cor}^{t}\left(\theta_{t}\right). \tag{5.38}$$

Intra-modal semantic discrimination learning: Beyond inter-modal semantic relevance learning, it is also essential to effectively learn discriminative features within each modality. Take haptic modality as an example, the objective is to enhance the similarity of z_t belonging to the same category, while reducing the similarity between projected features from different categories. The visual modality is identical. This is achieved by training a shared classifier. The specific loss function can be expressed as

$$L_{\text{dis}}\left(\theta_{l},\theta_{t},\theta_{c}\right)=-\frac{1}{N}\sum_{i=1}^{N}\left[y_{i}\left(\log p_{i}(z_{l}^{i})+\log p_{i}(z_{t}^{i})\right)\right], \tag{5.39}$$

where $p_i(z)$ denotes the probability distribution predicted by the classifier. y_i is the real label of the i-th sample, and θ_c represents the parameters of the shared classifier. By optimizing this loss function, we can respectively increase the difference among z_l within visual modality and z_t within haptic modality but belonging to different categories.

In summary, by jointly designing $L_{cor}(\theta_l, \theta_t)$ and $L_{dis}(\theta_l, \theta_t, \theta_c)$, z_l and z_t can be derived that possess both semantic discrimination and correlation.

5.5.4 Fusion for LR Visual and Haptic Modality Features

Following the acquisition of features about LR images and haptic signals, a power constraint is imposed [38]. Subsequently, these processed features are transmitted through a channel, which can be configured as either a Gaussian white noise channel or a Rayleigh channel. As a result, the noisy features z_{l-n} and z_{t-n} are obtained. To utilize these features for fitting the coding features of HR images and reconstructing HR images, multi-modal fusion is necessary, which can be supported by the GAN.

Specifically, z_{l-n} and z_{t-n} are input into the generative network. The fused feature z_m output by the generative network and the coding features z_h of the HR image are then used as inputs to the discriminator. Through adversarial training, which involves iteratively and continuously training the generator and discriminator, the loss L_m is optimized.

$$L_m = L_{m-adv}\left(G_m, D_m\right) + L_2\left(z_m, z_h\right). \tag{5.40}$$

The first item $L_{m\text{-adv}}\left(G_m, D_m\right)$ represents the standard GAN loss, which encourages the generative network to produce fused features that closely resemble the real HR image coding features. This loss can be expressed as

$$\min_{\theta_{gm}} \max_{\theta_{dm}} L_{m-\text{adv}}\left(G_m, D_m\right) = E_{z_h \sim p(z_h)} \log\left(D_m\left(z_h\right)\right) + E_{z_m \sim p(z_m)} \log\left(1 - D_m\left(z_m\right)\right). \tag{5.41}$$

The second term $L_2\left(z_m, z_h\right)$ is the L_2 loss that helps maintain semantic consistency.

5.5.5 Signal Reconstruction for HR Visual Signal

After determining the fused features, they are fed into the trained generator G_h for HR images. Following this, both G_h and the discriminator D_h are fine-tuned, which loss is designed as

$$L_{finet} = L_{per} + \beta L_{adv-finet} + \gamma L_{rec}, \tag{5.42}$$

where L_{per}, $L_{adv-finet}$, and L_{rec} denote the perceptual loss, adversarial fine-tuning loss, and reconstruction loss, respectively. The parameters $\boldsymbol{\beta}$ and $\boldsymbol{\gamma}$ represent the weights assigned to the different loss components.

Specifically, the perceptual loss has been empirically shown to enhance visual quality, as evidenced in [39]. It leverages a recognition network, such as VGG-19, which has been pre-trained on the ImageNet dataset, to generate feature maps. Instead of minimizing pixel-wise differences, it minimizes the discrepancy in the feature domain. It is defined as the L_2 loss between the feature maps extracted from both the real and reconstructed images,

$$L_{per} = \frac{1}{M_{i,j}} \left\| F^{i,j}(h) - F^{i,j}\left(G_h\left(z_m\right)\right) \right\|_2, \tag{5.43}$$

where $M_{i,j}$ denotes the number of parameters corresponding to the feature map. $F^{i,j}$ represents the output of the VGG-19 network after the ith convolutional layer and before the jth max-pooling layer. Here, $i = 4$, $j = 5$ are set, consistent with [39].

The adversarial fine-tuning loss $L_{adv-finet}$ can be expressed as

$$\min_{\theta_{gh}} \max_{\theta_{dh}} L_{adv-finet}\left(G_h, D_h\right) = -E_{z_m \sim p(z_m)}\left(D_h\left(G_h\left(z_m\right)\right)\right) + E_{h \sim p(h)}\left(D_h(h)\right), \tag{5.44}$$

where z_m indicates the fused features. θ_{gh} and θ_{dh} represent the parameters of the generator and discriminator, respectively.

Ultimately, the joint objective function for the feature mapping network, the feature fusion network, and the HR image generative network is formulated as follows:

$$L_{total} = L_{cor}\left(\theta_l, \theta_t\right) + L_{dis}\left(\theta_l, \theta_t, \theta_c\right) + L_m\left(\theta_{gm}, \theta_{dm}\right) + L_{finet}\left(\theta_{gh}, \theta_{dh}\right). \tag{5.45}$$

5.5.6 *Optimization Process*

Based on the preceding descriptions, the entire training optimization process can be categorized into two distinct phases. In the initial phase, the optimization focuses on the feature extraction network for HR images and the associated GAN. The parameters subject to optimization during this stage are $\left\{\theta_{eh}, \theta_{gh}, \theta_{dh}\right\}$. During this phase, the weights of a pre-trained large model are integrated to facilitate image feature extraction, thereby alleviating the training complexity of the model. The second phase centers on optimizing the feature mapping and fusion networks for LR images and haptic signals, as well as fine-tuning the subsequent GAN. The parameters to be optimized in this phase are $\{\theta_l, \theta_t, \theta_c, \theta_{gm}, \theta_{dm}, \theta_{gh}, \theta_{dh}\}$. Drawing on the work in [40], a strategy is implemented where each module of the model is trained individually before being combined for joint training. This enables flexible training and debugging of each module, facilitating modifications and adjustments to the network while mitigating the overall complexity of the training process. For these parameters, the Adam and RMSProp algorithms are employed for updating, as referenced in [41]. The entire optimization process is encapsulated in Algorithm 5.4. After a specific number of training steps, if the model's loss ceases to decrease, it signifies the convergence.

5.5.7 *Experimental Result and Analysis*

Experimental configuration

Dataset: The LMT surface material database is adopted here. In this database, the original image is first cropped to a dimension of 128 × 128, which serves as the HR image. Subsequently, the LR image is derived by applying bicubic down-sampling

Algorithm 5.4 Optimization process of the HaSR [34]

Require: Database $D = \{d_i\}_{i=1}^{N}$, where d_i consists of a HR image h_i, the corresponding LR image l_i and haptic signal t_i;
Initial parameters $\theta_{eh}, \theta_{gh}, \theta_{dh}, \theta_l, \theta_t, \theta_c, \theta_{gm}, \theta_{dm}$;
Learning rate μ_1, μ_2, iterations n_1, n_2, batch size.

Stage 1: Encoding and decoding of HR images
for iteration $= 1, \ldots, n_1$ **do**
  Update $\theta_{eh}, \theta_{dh}, \theta_{gh}$ by RMSProp for $L_{pre-adv}$ and L_{pre}
  Update $\theta_{eh} \leftarrow \theta_{eh} - \mu_1 \nabla_{\theta_{eh}} L_{pre}$
  Update $\theta_{dh} \leftarrow \theta_{dh} - \mu_1 \nabla_{\theta_{dh}} L_{pre-adv}$
  Update $\theta_{gh} \leftarrow \theta_{gh} - \mu_1 \nabla_{\theta_{gh}} L_{\mathrm{pre}}$
end for
return $E_h(\theta_{eh}), G_h(\theta_{gh}), D_h(\theta_{dh})$
Stage 2: Optimizing feature mapping network, fusion network, and fine-tuning of HR image generation network
for iteration $= 1, \ldots, n_2$ **do**
  Update $\theta_c, \theta_l, \theta_t$ by Adam for L_{cor} and L_{dis}
  $\theta_c \leftarrow \theta_c - \mu_2 \nabla_{\theta_c} L_{dis}$
  $\theta_l \leftarrow \theta_l - \mu_2 \nabla_{\theta_l} (L_{cor} + L_{dis})$
  $\theta_t \leftarrow \theta_t - \mu_2 \nabla_{\theta_t} (L_{cor} + L_{dis})$
  Update θ_{dm} and θ_{gm} by Adam for L_m
  $\theta_{gm} \leftarrow \theta_{gm} - \mu_2 \nabla_{\theta_{gm}} L_m$
  $\theta_{dm} \leftarrow \theta_{dm} - \mu_2 \nabla_{\theta_{dm}} L_{m-adv}$
  Update θ_{gh} and θ_{dh} by RMSProp for L_{finet}
  $\theta_{gh} \leftarrow \theta_{gh} - \mu_2 \nabla_{\theta_{gh}} L_{\mathrm{finet}}$
  $\theta_{dh} \leftarrow \theta_{dh} - \mu_2 \nabla_{\theta_{dh}} L_{adv-\mathrm{finet}}$
end for
return $F_l(\theta_l), F_t(\theta_t), G_m(\theta_{gm}), D_m(\theta_{dm}), G_h(\theta_{gh}), D_h(\theta_{dh})$

to the HR image, yielding an image size of 32 × 32. In the context of haptic data, given the impact of pressure exerted during the contact and release phases of the haptic stylus, the vibration signal captured during the stable middle contact period is extracted and designated as the final haptic signal. Following this, the corresponding spectrogram is generated using the short-time Fourier transform algorithm, configured with a 32-point Hamming window and an 8-point hop size. Ultimately, for each material type, a dataset comprising 20 HR image samples, 20 LR image samples, and 20 haptic samples is assembled. Within the database, 80% of the samples are randomly selected for training purposes, with the remaining.

Competing schemes: To assess the visual signal recovery performance of HaSR framework, a comparative analysis is conducted with seven competing schemes. Pix2Pix [23] is a seminal approach that leverages image-conditional GAN for supervised image-to-image translation. In this study, the spectrogram of the haptic signal is utilized as conditional information to generate HR images. DiscoGAN [24] employs GANs to learn and uncover the relationships between different domains. It converts spectrogram of the haptic signal to HR images. Bicubic interpolation is a widely adopted image upsampling technique that achieves clearer images by directly interpolating low-resolution ones. SRGAN [42] integrates GAN into the realm of super-resolution reconstruction. ESRGAN [39] builds upon SRGAN by introducing a residual-in-residual dense block architecture. This approach enhances visual quality by balancing perceptual quality and fidelity. NLSA [43] is a nonlocal sparse attention method that identifies the most informative locations globally while disregarding irrelevant regions. SR3 [44] adapts the denoising diffusion probabilistic model to perform super-resolution reconstruction via a stochastic iterative denoising process.

Implementation details: The implementation details of all the modules and networks in the HaSR are listed in the following.

HR image encoder: The HR image encoders network E_h consists of a VGG-16 network followed by two fully connected layers with dimensions of 512 and 128, respectively. The resultant 128-dimensional feature vector serves as the encoded representation of the HR image.

Generator and discriminator for HR Image: The generator G_h for HR images is composed of fully connected layers and deconvolutional networks. The convolutional kernels have a size of 3×3, with the corresponding number of filters being 512, 256, 128, 64, and 3, respectively. The discriminator D_h is constructed using convolutional networks and fully connected layers. In the convolutional networks, the kernel size is 3×3, and the number of filters is 64, 128, 256, and 512, respectively. The dimensions of the fully connected layers are 512, 128, and 1, respectively.

Encoders for haptic Signal and LR Image: The haptic encoding network E_t comprises a DenseNet-121 network and fully connected layers with dimensions of 1024 and 256. The LR image encoding network E_l is analogous to the HR image encoding network, with the exception that the input image size is $32 \times 32 \times 3$. The 256-dimensional feature vector generated by this network serves as the encoded representation of the LR images.

Feature mapping network: Two two-layer fully connected layers with dimensions of 256-128 are employed to project the encoded features. Additionally, a common classifier is constructed using three fully connected layers with dimensions of 128, 32, and 9, respectively.

Feature fusion network: The generative network G_m firstly concatenates the projected features of the LR images and the haptic signals, and then processes them through fully connected layers with dimensions of 256 and 128. The discriminator D_m is composed of fully connected layers with dimensions of 128, 64, and 1.

Evaluation metrics: Three metrics are selected for evaluating performance of the HaSR. Specifically, the peak signal-to-noise ratio (PSNR) is used to quantify the similarity between pixel values of reconstructed and ground-truth images. SSIM serves as a comprehensive measure of image similarity, incorporating three critical components: luminance, contrast, and structural information. Fréchet inception distance (FID) is to assess the performance of GAN. It quantifies the discrepancy between the distributions of real and generated images by leveraging the Fréchet distance in the feature space of the Inception network.

Results and analysis

Table 5.11 Quantitative results of the HaSR and the other competing approaches

Approach	PSNR	SSIM	FID
Pix2Pix	18.54	0.344	220.527
DiscoGAN	25.75	0.643	170.249
BicuBic	27.14	0.684	247.741
SRGAN	28.12	0.792	88.478
ESRGAN	28.46	0.824	67.747
NLSA	29.77	0.879	58.451
SR3	32.11	0.912	34.893
HaSR	**30.88**	**0.897**	**31.256**

Here, a comprehensive evaluation of the HaSR approach, encompassing both quantitative and qualitative analyses is performed. Table 5.11 presents the mean values of PSNR, SSIM, and FID for the generated images across various methods. Specifically, HaSR significantly outperforms the other methods in all three metrics. This is primarily due to the impracticality of geometric alignment between input and output in our context, which is a requirement for these cross-domain methods. Compared to SRGAN and ESRGAN, HaSR achieves improvement in FID. This enhancement is attributed to the incorporation of haptic signals and the exploration of semantic correlations between different modalities in HaSR. The results of NLSA are comparable to HaSR in terms of PSNR and SSIM, which can be ascribed to NLSA's ability to integrate nonlocal operations with sparse coding. SR3 attains optimal results on these metrics by leveraging the latest diffusion probabilistic model. However, this comes at the cost of increased computation time. Additionally, Table 5.12 demonstrates that the HaSR exhibits greater stability compared to NLSA and SR3, as evidenced by the lower standard deviation of multiple experiments. This enhanced stability is owing to the effective training strategy.

Secondly, Fig. 5.20 illustrates a comparative series of HR images synthesized by the methods previously described, thereby facilitating a qualitative evaluation of their respective performance characteristics. The HaSR approach is found to exhibit markedly superior reconstruction capabilities when compared with other competing methods. This superiority is evidenced by its ability to generate images that closely emulate the characteristics of real HR images across multiple dimensions, including texture, outline, color, and detailed information. Conversely, images generated by

Table 5.12 Standard deviation of multiple experiments for the HaSR and the other competing approaches

Standard deviation	NLSA	SR3	HaSR
PSNR	3.156	4.589	**2.564**
SSIM	0.078	0.091	**0.062**

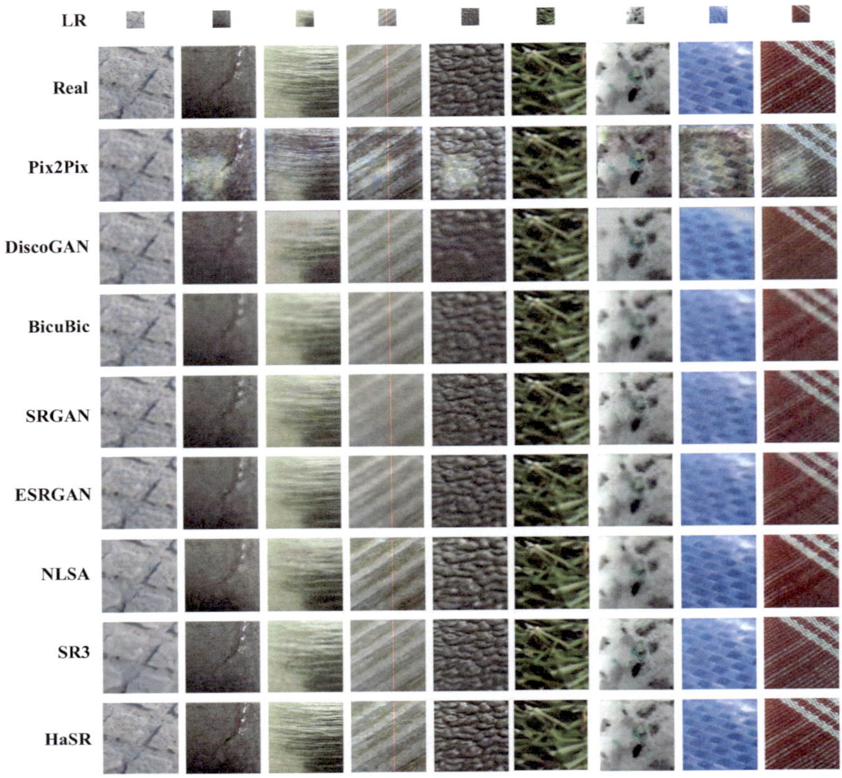

Fig. 5.20 The reconstructed HR images by the HaSR and the other competing approaches

the Pix2Pix method are observed to exhibit significant noise artifacts, which detract from their visual fidelity. Similarly, the images produced by DiscoGAN and the Bicubic interpolation method suffer from pronounced blurriness, indicating limitations in their ability to accurately reconstruct fine details. While the image qualities of the SRGAN and the ESRGAN have shown improvement to some extent, they still exhibit deficiencies in detailed information when compared to real images. Furthermore, although the images generated by the NLSA method contain a higher degree of detailed information, they are compromised by the presence of artifacts and blurring in the corners of the images. The SR3 method, through its iterative refinement process, achieves a more robust distribution of image features, resulting in qualitative performance that is comparable to that of the HaSR approach.

In order to systematically investigate the efficacy of individual components within the HaSR, a comparative analysis is conducted between the full HaSR and several ablated variants. Specifically, three distinct schemes are examined. Specifically, to assess the contribution of haptic information, a variant is designed that solely utilizes the coding features extracted from LR images to approximate the coding features

Table 5.13 Results about impacts of components in the HaSR

Approach	PSNR	SSIM	FID
HaSR w/o haptic	26.47	0.659	126.527
HaSR w/o modal constraints	27.42	0.783	94.184
HaSR w/o our fusion (just concatenation)	28.75	0.856	80.229
HaSR (fully)	**30.88**	**0.897**	**31.256**

of HR images, designated as "HaSR w/o haptic". To evaluate the necessity of the feature mapping networks, a second variant is devised that omits the constraints of inter-modal semantic consistency and intra-modal discrimination. Instead, the coding features of both haptic signals and LR images are directly fed into the fusion network without these constraints. This scheme is referred to as "HaSR w/o modal constraints". To scrutinize the effectiveness of the fusion strategy employed in HaSR, a third variant is implemented that employs a simple concatenation method to combine the projected features, followed by a fully connected layer to align the feature dimensions, labeled as "HaSR w/o modal fusion". Table 5.13 presents the PSNR, SSIM, and FID values of the images generated by the full HaSR and the three aforementioned ablated variants. The results indicate that the images produced by the full HaSR consistently outperform those generated by the other three variants, thereby underscoring the significance of integrating haptic signals, utilizing the feature mapping networks, and employing the advanced fusion strategy. Notably, the variant lacking haptic signals demonstrated markedly inferior performance, which further substantiates the critical role of haptic information in enhancing the super-resolution reconstruction process.

Finally, in Fig. 5.21, the influence of channel bandwidth and SNR on the performance of the HaSR approach is investigated. Specifically, experiments are conducted in both AWGN channels and Rayleigh fading channels, as opposed to ideal channel conditions characterized by zero noise power. It should be noted that a distinct network is trained for each combination of SNR and channel bandwidth.

Given a channel input $x \in \mathbb{R}^B$, where B is the channel bandwidth, an average power constraint of $P = 1$ is imposed on every channel input vector, such that $1/B \sum_{i=1}^{B} |x_i|^2 \leq P = 1$. To accommodate different bandwidth settings, the dimensions of the fully connected layers in the encoding and mapping networks are adjusted accordingly. Figure 5.21 illustrates the PSNR achieved under various channel and bandwidth conditions as a function of SNR. As SNR increases, the quality of the reconstructed images improves monotonically. However, it is observed that different channels exhibit distinct performance characteristics when the model reaches a steady state. Specifically, the HaSR can approach the noiseless bound under AWGN channels with sufficiently high SNR values. In contrast, for Rayleigh fading channels, even with increased bandwidth and SNR, the HaSR method fails to achieve the PSNR values attainable under ideal conditions. These results indicate that the HaSR

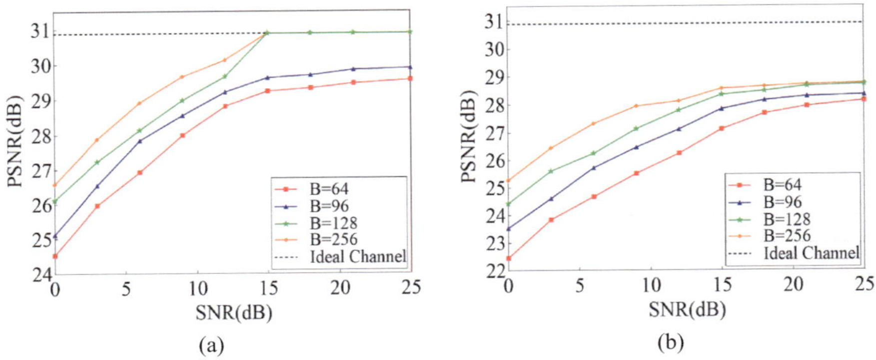

Fig. 5.21 Performance of the HaSR under different channels, SNRs, and bandwidth conditions: **a** AWGN channel; **b** Rayleigh fading channel

provides a certain degree of denoising capability, thereby enhancing the quality of the reconstructed HR images. However, it is unable to completely mitigate the effects of variable channel gain on the channel symbols in Rayleigh fading channels. Additionally, it is observed that reconstruction quality and model robustness improve with increasing bandwidth, while larger bandwidths are more prone to reaching saturation at lower SNR levels.

5.6 Summary

In this chapter, we have described cross-modal signal recovery techniques in the aspects of whole structure, modules, optimization process, experimental result and analysis. Firstly, according to the categories of multi-modal services, the corresponding recovery approaches have been respectively designed. For the haptic-dominant multi-modal services, an audio-visual-aided haptic signal recovery scheme is proposed, while for the audio-visual-dominant multi-modal services, a haptic-aided visual signal recovery scheme is presented. By employing steps such as feature extraction, modality fusion, and desired signal recovery, the received quality of lost, damaged, or delayed arrival of signals can be guaranteed. In particular, for the equally important multi-modal services, a visual-haptic mutual signal recovery scheme is provided, which effectively relaxes the assumption of the two schemes mentioned above. It not only explores and transfers knowledge from the auxiliary databases, but also realizes high quality visual and haptic signal recovery by utilizing both coarse-grained and fine-grained recovery techniques. Although there exists a little performance gap compared with the former two approaches, it can effectively handle challenges about lost, damaged, or delayed arrival of both visual and haptic signals. Secondly, as some valuable contents may be received or cached at the receiver or master domain in advance, information retrieval becomes another efficient and convenient tool for supporting signal recovery. Based on this, intra-modal one-to-one retrieval, inter-modal one-to-one retrieval, and inter-modal one-to-many retrieval ways are designed for solving the encountered problems from different scenarios. Experimental results show the performance advantages and robustness of this scheme. Finally, to deal with the issue about constrained bandwidth and the presence of modality competition in applications such as remote industrial robotic grasping, only low-resolution visual signals and haptic signals are transmitted while the desired high-resolution visual signals can be derived at the receiver. According to this demand, a cross-modal super-resolution signal reconstruction approach is presented. The experimental results validate the reconstruction quality and model robustness under different communication environments.

References

1. Wei X, Shi Y, Zhou L (2022) Haptic signal reconstruction for cross-modal communications. IEEE Trans Multimed 24:4514–4525
2. Arandjelovic R, Zisserman A (2017) Look, listen and learn. In: Proceedings of the IEEE international conference computer vision, Hawaii, USA, July 21–26, pp 609–617
3. Baltrušaitis T, Ahuja C, Morency LP (2019) Multimodal machine learning: a survey and taxonomy. IEEE Trans Pattern Anal Mach Intell 41(2):423–443
4. He Y, Xiang S, Kang C, Wang J, Pan C (2016) Cross-modal retrieval via deep and bidirectional representation learning. IEEE Trans Multimedi 18(7):1363–1377
5. Gemmeke JF, Ellis DP, Freedman D, Jansen A, Lawrence W, Moore RC, Plakal M, Ritter M (2017) Audio set: an ontology and human-labeled dataset for audio events. In: Proceedings of the IEEE international conference on acoustics, speech and signal processing, New Orleans, USA, Mar 5–9, pp 776–780
6. Strese M, Schuwerk C, Iepure A, Steinbach, (2017) Multimodal feature based surface material classification. IEEE Trans Haptics 10(2):226–239
7. Purri M, Dana K (2020) Teaching cameras to feel: estimating tactile physical properties of surfaces from images. In: Proceedings of the European conference on computer vision, online, Aug 23–28, pp 1–20
8. Takahashi K, Tan J (2019) Deep visuo-tactile learning: estimation of tactile properties from images. In: Proceedings IEEE International Conference on Robotics and Automation, Montreal, Canada, May 20–24, pp 8951–8957
9. Liu H, Guo D, Zhang X, Zhu W, Fang B, Sun F (2021) Toward image-to-tactile cross-modal perception for visually impaired people. IEEE Trans Autom Sci Eng 18(2):521–529
10. Wei X, Zhang M, Zhou L (2022) Cross-modal transmission strategy. IEEE Trans Circ Syst Vid 32(6):3991–4003
11. Liu H, Wang F, Sun F, Fang B (2019) Surface material retrieval using weakly paired cross-modal learning. IEEE Trans Autom Sci Eng 16(2):781–791
12. Zheng W, Liu H, Wang B, Sun F (2019) Cross-modal surface material retrieval using discriminant adversarial learning. IEEE Trans Ind Informat 15(9):4978–4987
13. Vaswani A, Shazeer N, Parmar N, Uszkoreit J, Jones L, Gomez A N, Kaiser L, Polosukhin I (2017) Attention is all you need. In: Proceedings of the advances in neural information processing systems, Long Beach, USA, Dec 4–9, pp 1–15
14. Goodfellow I, Bengio Y, Courville A (2016) Deep learning. MIT Press, Cambridge, MA, USA
15. Zhou Y, Liu H, Li D, Cao H, Yang J, Li Z (2019) Cross-sensor image change detection based on deep canonically correlated autoencoders. In: Proceedings of the artificial intelligence for communications and networks, Online, July 5, pp 251–257

16. Wang CH, Chuang SP, Lee HY (2019) Towards audio to scene image synthesis using generative adversarial network. In: Proceedings of the IEEE international conference on acoustics speech signal processing, Brighton, UK, May 12–17, pp 496–500
17. Jin Q, Lin R, Yang F (2020) E-WACGAN: enhanced generative model of signaling data based on WGAN-GP and ACGAN. IEEE Syst J 14(3):3289–3300
18. Chen Y, Wu G, Zhou J, Qi G (2019) Image generation via latent space learning using improved combination. Neurocomputing 340:8–18
19. Wei X, Liao J, Zhou L, Sari H, Zhuang W (2024) Toward generic cross-modal transmission strategy. IEEE Trans Commun 72(10):6059–6072
20. Zhuang F, Qi Z, Duan K, Xi D, Zhu Y, Zhu H, Xiong H, He Q (2021) A comprehensive survey on transfer learning. Proc IEEE 109(1):43–76
21. Dana KJ, Ginneken BV, Nayar SK, Koenderink JJ (1997) Columbia-Utrecht reflectance and texture database. http://www.cs.columbia.edu/CAVE/curet
22. Dai D, Riemenschneider H, Gool LV (2014) The synthesizability of texture examples. In: Proceedings of the IEEE conference on computer vision and pattern Recognition, Columbus, USA, June 23–28, pp 3027–3034
23. Isola P, Zhu J Y, Zhou T, Efros AA (2017) Image-to-image translation with conditional adversarial networks. In: Proceedings of the IEEE conference on computer vision and pattern Recognition, Honolulu, USA, July 21–26, pp 1125–1134
24. Kim T, Cha M, Kim H, Lee JK, Kim J (2017) Learning to discover cross-domain relations with generative adversarial networks. In: Proceedings of the international conference on machine learning, Sydney, Australia, Aug 6–11, pp 1857–1865
25. Hao W, Zhang Z, Guan H (2018) CMCGAN: a uniform framework for cross-modal visual-audio mutual generation. In: Proceedings of the AAAI conference on artificial intelligence, New Orleans, USA, Feb 2–7, pp 6886–6893
26. Zhou Y, Liu H, Li D, Cao H, Yang J, Li Z (2019) Cross sensor image change detection based on deep canonically correlated autoencoders. In: Proceedings of the international conference on artificial intelligence for communications and networks, Harbin, China, Apr 16–18, pp 251–257
27. Takahashi K, Tan J (2019) Deep visuo-tactile learning: estimation of tactile properties from images. In: Proceedings of international conference on robotics and automation, Montreal, Canada, May 20–24, pp 8951–8957
28. Griffin D, Lim J (1984) Signal estimation from modified short-time Fourier transform. IEEE Trans. Acoust., Speech, Signal Process. ASSP-32(2):236–243
29. Zhou W, Bovik AC, Sheikh HR, Simoncelli EP (2004) Image quality assessment: from error visibility to structural similarity. IEEE Trans Image Process 13:600–612
30. Hassen R, Steinbach E (2020) Subjective evaluation of the spectral temporal SIMilarity (ST-SIM) measure for vibrotactile quality assessment. IEEE Trans Haptics 13(1):25–31
31. Aoudia FA, Hoydis J (2019) Model-free training of end-to-end communication systems. IEEE J Sel Areas Commun 37(1):2503–2516
32. Xu J, Wei X, Zhou L (2022) Information recovery technology for cross-modal communications. Acta Electron Sin 50(7):1631–1642
33. Wang Z, Chen J, Hoi SCH (2021) Deep learning for image super resolution: a survey. IEEE Trans Pattern Anal Mach Intell 43(10):3365–3387
34. Liu H, Wei X, Gao Y, Zhou L, Qian Y (2024) Super-resolution reconstruction for cross-modal communications in industrial internet of things. IEEE Internet Things 11(13):24148–24161
35. Girbes-Juan V, Schettino V, Demiris Y, Tornero J (2021) Haptic and visual feedback assistance for dual-arm robot teleoperation in surface conditioning tasks. IEEE Trans Haptics 14(1):44–56
36. Luan X, Zheng J, Li W (2022) Learning unsupervised face normalization through frontal view reconstruction. IEEE Trans Circ Syst Vid 32(8):5201–5212
37. Arjovsky M, Chintala S, Bottou L (2017) Wasserstein generative adversarial networks. In: Proceedings of the international conference on machine learning, Sydney, Australia, Aug 6–11, pp 214–223

38. Kurka DB, Gündüz D (2021) Bandwidth-agile image transmission with deep joint source-channel coding. IEEE Trans Wireless Commun 20(12):8081–8095
39. Wang X, Yu K, Wu S, Gu J, Liu Y, Dong C, Qiao Y, Loy CC (2018) ESRGAN: enhanced super-resolution generative adversarial networks. In: Proceedings of the European conference on computer vision, Munich, Germany, Sep 8–14, pp 1–6
40. Jankowski M, Gündüz D, Mikolajczyk K (2021) Wireless image retrieval at the edge. IEEE J Sel Areas Commun 39(1):89–100
41. Cai S, Zhu K, Ban Y, Narumi T (2021) Visual-tactile cross-modal data generation using residue-fusion GAN with feature-matching and perceptual losses. IEEE Robot Autom Lett 6(4):7525–7532
42. Ledig C, Theis L, Huszar F, Caballero J, Cunningham A, Acosta A, Aitken A, Tejani A, Totz J, Wang Z, Shi W (2017) ESRGAN: enhanced super-resolution generative adversarial networks. In: Proceedings of the IEEE conference on computer vision and pattern recognition, Honolulu, USA, July 21–26, pp 4861–4890
43. Mei Y, Fan Y, Zhou Y (2021) Image super-resolution with non-local sparse attention. In: Proceedings of the IEEE conference on computer vision and pattern recognition, Nashville, USA, June 19–25, pp 3517–3526
44. Saharia C, Ho J, Chan W, Salimans T, Fleet DJ, Norouzi M (2023) Image super-resolution via iterative refinement. IEEE Trans Pattern Anal Mach Intell 45(4):4713–4726

Chapter 6
Cross-Modal Communication Prototype System

Abstract Based on theoretical and technical investigations, in this chapter, several developed cross-modal communication prototype systems in the areas of industrial manipulation, smart education, and intelligent healthcare have been described. Firstly, a large-scale multimodal database named VisTouch has been established, which supports the research of cross-modal communications. Secondly, the details about the design and implementation of a visual-haptic human-machine interactive system are provided, which effectively meets the demands of the remote industrial robotic grasping applications. Thirdly, a virtual acupuncture skill training application is constructed to support the emerging remote practical teaching demands. Finally, a remote throat swab sampling platform is constructed, which is of great significance in protecting the lives of medical staff and breaking the chains of virus transmission during the COVID-19 pandemic.

6.1 Large-Scale Multi-Modal Database

To address the research requirements of cross-modal communications, a large-scale multi-modal (audio, visual, and haptic) dataset named VisTouch is constructed (http://8.133.175.194). This section mainly describes the data collection process of the VisTouch, introducing the audio, visual, and haptic acquisition devices as well as the types of materials [1].

6.1.1 Multi-Modal Data Collection

Haptic perception is related to the properties of the contacted object and the manner in which the surface is explored, with friction playing a significant role. To this end, in the VisTouch, the data collection method involves a script-controlled robotic arm sliding to touch various materials. The sliding frictional force generated between the fingertip and the material during the sliding touch process is used as the haptic signal. Meanwhile, a high-definition camera and a unidirectional microphone

X. Wei et al., *Cross-Modal Communication Technology*, Wireless Networks,
https://doi.org/10.1007/978-3-032-00957-9_6

mounted on the robotic arm are employed to capture audio and visual signals, which are synchronized using time stamps.

Accurate and low-noise acquisition of haptic signals is the core of VisTouch construction. The magnitude of the sliding frictional force is related to the normal pressure applied to the contact surface and the kinetic friction coefficient, which reflects the material properties and is generally a constant value. Therefore, applying a constant normal pressure is key to ensuring the accuracy and low noise of the haptic signal. This requires a two-pronged approach. On one hand, we place the robotic arm on the table and provide a constant downward force perpendicular to the table surface to the robotic hand mounted at the end of the arm. On the other hand, we select materials in sheet form to ensure the normality of the driving force to the contact surface, thereby reducing the impact of material shape factors on the collected signals.

To enhance the sample diversity and practicality of the dataset, two data augmentation methods are introduced during the data collection process. One is setting three types of sliding touch trajectories: straight-line sliding, curved sliding, and zigzag sliding. The other is setting three levels of constant normal driving force (3 newtons, 6 newtons, 9 newtons), which are cross-combined with the sliding trajectories, resulting in a total of nine sliding methods.

6.1.2 Signal Sample in the VisTouch Dataset

The characteristics of a signal are closely related to the material itself. When a robotic arm is controlled to touch various materials, the haptic signal curve of rough materials (such as "stone") fluctuates more than that of smooth materials (such as "glass"), and the sound signal is harsher. Therefore, in theory, the more material samples included in the dataset, the more it helps to explore the mechanisms and representation models of audio, visual, and haptic perception. The shape and material of the samples affect the form and content of multi-modal signals. Additionally, the haptic acquisition of the sample shape requires an array of dot matrix sensors. Therefore, we use only sheet-like material samples (such as stone slices, paper slices, and wood slices) for data acquisition here. We use a single force sensor for this purpose. The acquired haptic signals are then represented as one-dimensional time series. In the future, it will be considered to introduce new types of sensors, such as tactile gloves or GelSight, to perceive the shape information of objects and form two-dimensional tactile maps, thereby enriching the content of the dataset and enhancing its application value. Regarding the sample types, 47 common and highly practical materials in current life have been surveyed and classified, as shown in Table 6.1. Then, multi-modal signal acquisition is performed on these material samples using the acquisition method designed in Sect. 6.1.1. An example of the VisTouch dataset is shown in Fig. 6.1.

Finally, the constructed VisTouch dataset is compared with existing mainstream datasets in terms of modality, category, and sample quantity, as shown in Table 6.2. The frame quantity refers to the number of frames of the collected image data. For

Table 6.1 Sample category containing in the VisTouch dataset

Category	Subcategory	Kinds
Plastic	Polyethylene terephthalate, high-density polyethylene, polyvinyl chloride, low-density polyethylene, polypropylene, polystyrene	6
Metal	Iron, aluminum, nickel, zinc, titanium, indium, copper, tantalum, molybdenum	9
Wood	Cherry wood, pine, black walnut, bamboo	4
Paper	Printer paper, newspaper, cardboard	3
Ceramics	Traditional ceramics, special ceramics	2
Rubber	Natural rubber, synthetic rubber	2
Natural textiles	Cotton, linen, silk	3
Synthetic textiles	Nylon, polyester, acrylic, vinylon, polypropylene, spandex, carbon fiber	7
Glass	Soda-lime glass, quartz glass	2
Leather	Cowhide, sheepskin, artificial leather	3
Stone	Granite, marble, limestone, slate, shale, andesite	6

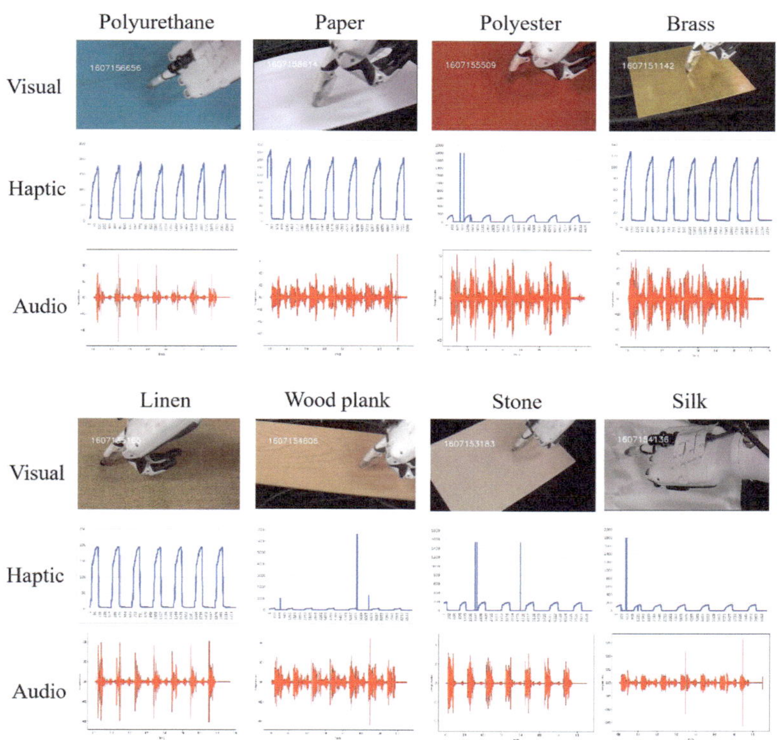

Fig. 6.1 Illustration of the signal samples in the VisTouch dataset

Table 6.2 The comparison of VisTouch and the other representative datasets

Name	Modality	Kinds	Scale
URMP [2]	Audio, visual	14	Million
VisGel [3]	Visual, haptic	195	Million
STAG [4]	Haptic	26	Hundred-thousand
VisTouch	Audio, visual, haptic	47	Ten-million

single-modal haptic data such as STAG, it refers to the number of haptic map frames. The datasets related to touch mainly include VisGel and STAG. Compared with them, the differences of the VisTouch dataset mainly lie in two main aspects. On one hand, it contains three-modality signals (audio, visual, and haptic signals), while VisGel only records visual and haptic signals, and STAG only records haptic signals. On the other hand, the haptic signals of VisGel and STAG are essentially mechanical pressures, which are only related to the shape of the object and the applied force, and cannot accurately reflect the material properties. In contrast, the VisTouch dataset uses sliding friction force as the haptic signal, and its kinetic friction coefficient is related to the material itself, so it can accurately reflect the material properties.

6.2 Visual-Haptic Human-Machine Interactive System

Based on the key techniques described in Sects. 3–5, a visual-haptic human-machine interactive system can be constructed, shown in Fig. 6.2. Here, a glove with haptic perception has been developed using small-sized, high-torque linear servo drives. It can follow finger movements to read data in a passive mode and simulate force haptic feedback in an active mode. Then, the arm movement data of the operator is obtained using the developed glove and transmitted via a wireless channel to the Aubo-i3 robotic arm to form movement instructions. According to these instructions, the Aubo-i3 robotic arm controls the Inspire robotic hand to grasp objects in accordance with the operator's gestures, while the force information of the Inspire robotic hand is fed back to the glove, enabling the operator to simultaneously perceive real haptic feedback. In addition, the cross-modal signal recovery method proposed in Sect. 5.1 has been loaded. It uses the visual signals collected by the Kinect depth camera to compensate for the loss of haptic signals caused by remote transmission, mechanical vibration, and other reasons in the above process, thereby enhancing the stability of human-machine interaction.

When considering the development process in detail, there exist two key steps, shown in Fig. 6.3. One is the "motion-following control". In this step, the motion capturing module acquires the operator's movement data at the master domain, identifies the outline of the operator's arm, and calculates the arm's position, velocity, and direction using a computer vision technique. The data is then converted into control

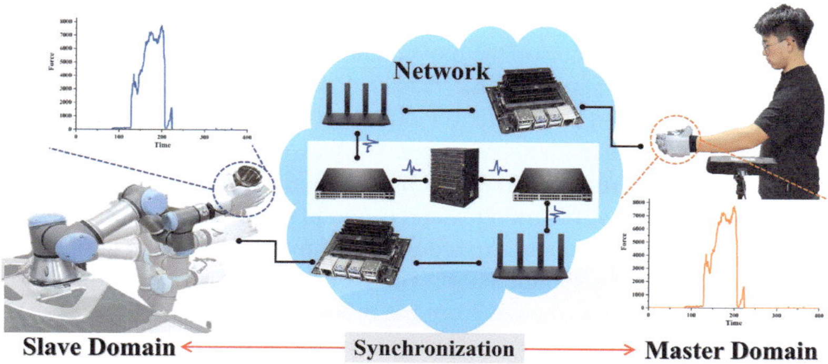

Fig. 6.2 Illustration of the visual-haptic human-machine interactive system

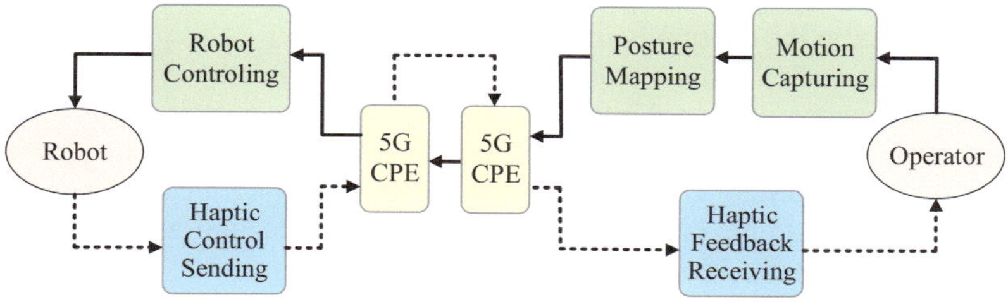

Fig. 6.3 Work flow of the interactive system

commands for the robotic arm through a posture mapping module. Meanwhile, the haptic feedback glove reads the movement data of the operator's hand and converts it into control commands for the robotic hand. These commands are sent to the slave domain via a 5G CPE, cached by the robot controller, and then used to control the robotic arm to follow the operator's movements. The other is "haptic feedback". When the robotic hand at the slave domain interacts with an object, it collects force data from each finger. This data is transmitted via 5G to the master domain, where it is mapped into control commands for haptic feedback. These commands drive the linear motors on the haptic feedback glove to move the metal rods, thereby providing haptic feedback to the operator's hand.

Recently, this interactive system has been deployed on the inspection robot of a certain company to achieve remote scanning and quality inspection. The original inspection robot contains an AGV vehicle, a high-definition pan-tilt camera, and a robotic arm which operates in the workshop. During the practical deployment, there are several key technical improvements that need attention.

(1) Adjusting the camera position: At the workshop (slave domain), the high-definition pan-tilt camera mounted at the end of the robotic arm should be moved to the rear of the arm. The installation position is adjusted so that it can fully capture the robotic arm, the robotic hand, and the operating table. The stability

of the camera is also adjusted to prevent it from shaking significantly due to inertia or other factors during the movement of the vehicle. This could affect the observation by the operator at the master domain. Additionally, the camera is calibrated for position and angle to facilitate subsequent observation.

(2) Adding the robotic hand and controller: At the workshop (slave domain), a six-degree-of-freedom robotic dexterous hand is installed at the end of the robotic arm. The data cable of the robotic hand is bound to the robotic arm in a way that does not hinder its movement. An industrial computer is installed inside the AGV vehicle to control the robotic hand. The industrial computer sends commands and receives status information from the robotic hand via wired transmission.

(3) Adding the exoskeleton glove: At the exhibition booth (master domain), a haptic feedback exoskeleton glove is added for the operator. Using 3D printing, the operator's fingers are modeled, and new finger covers are printed to fit the ergonomic design, providing the operator with more realistic haptic feedback and enhancing the operability of the entire system.

(4) Adding the motion capture module: At the exhibition booth (master domain), a binocular camera and an AI-based motion capturing algorithm are added. The binocular camera acquires depth information and key features of the operator's arm. An AI algorithm converts this information into control commands, which are transmitted to the robotic arm via a 5G communication module, enabling contactless human optical motion capture.

(5) Adding the 5G communication module: 5G CPEs are installed on both the exhibition booth and the AGV vehicle in the workshop to achieve efficient cross-modal transmission. This includes control information for the robotic arm, high-definition image information from the binocular camera, haptic information from the six-degree-of-freedom robotic dexterous hand, and videos between the master and slave domains.

(6) Adding the teleoperation system: Using the 5G communication module as a bridge, all software and hardware resources are integrated to build the control platform of the teleoperation system. This enables more precise information acquisition and more intuitive feedback, improving human-machine interaction efficiency.

(7) Setting up the operating table: In the workshop, an operating table of appropriate height is set up. An image output device is placed on the table to provide a scanning interaction interface, and the barcode scanner is positioned within the safe working range of the robotic arm.

After finishing deployment, the operator uses the robotic arm to complete the scanning quality inspection. Unlike existing remote teleoperation systems, this system introduces cross-modal coding, transmission, and signal recovery techniques, adding more intuitive visual, touch, and auditory guidance for the operator. The operator, wearing a haptic feedback exoskeleton glove, can precisely and stably drive the robot to move, grasp the barcode scanner, and complete the scanning quality inspection with the help of motion capture, visual feedback, force feedback, and 5G communication. This enhances the efficiency and performance of human-machine interaction.

6.3 Virtual Acupuncture Skill Training Application

At present, the implementation of remote practical teaching remains cumbersome, particularly for courses that necessitate hands-on operations or laboratory experiments. To address this challenge, we select acupuncture skill training as a case study and develop a virtual interactive application leveraging cross-modal communication technology, shown in Fig. 6.4. This application comprises three main components: vivid haptic rendering, augmented reality (AR), and a skill assessment subsystem. Notably, the vivid haptic rendering encompasses a cross-modal haptic recovery algorithm designed to synthesize haptic force feedback. It also deploys a self-assembled touch sensation perception device (a Geomagic Touch connected to an acupuncture needle) that renders this haptic force feedback. AR provides a virtual interactive training environment displayed in AR devices. The skill assessment subsystem evaluates students' manipulation skills by comparing the needle motion and haptic signals from their training manipulation with those of the standard manipulation across four dimensions: position, velocity, acceleration, and force. After a semester of utilization, this system has proven effective in vividly conveying sensations such as "the

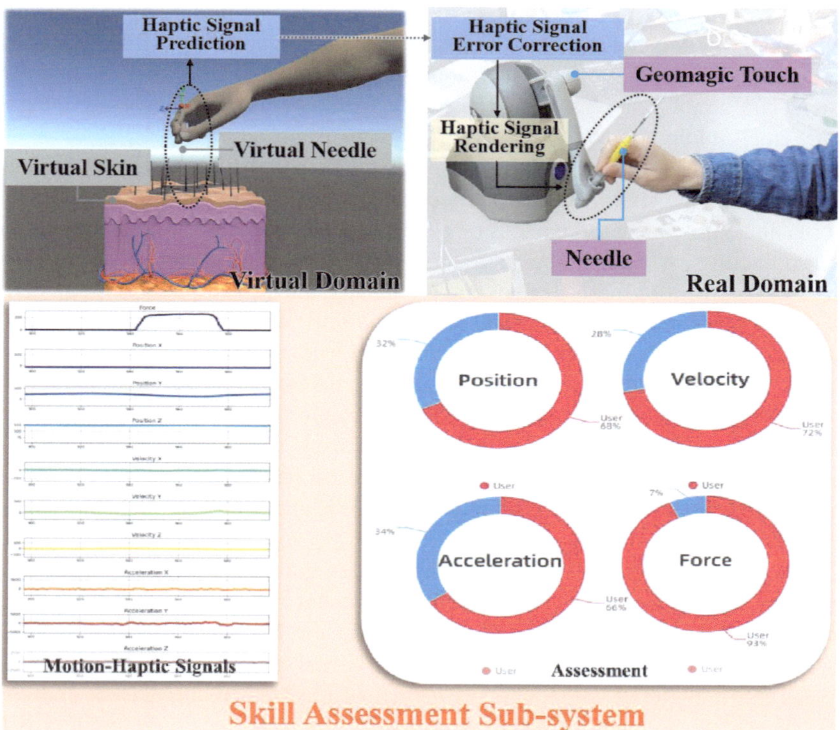

Fig. 6.4 The structure of virtual acupuncture skill training application

qi" and "needling", thereby correcting and enhancing students' acupuncture skills. The details of this application are described in the following.

6.3.1 Hardware and Software

Hardware: Unconstrained and advanced hardware is crucial for delivering an immersive experience. Initially, the selection of augmented reality (AR) devices and haptic devices must be considered. Regarding AR devices, AR enables students to visualize and interact with the virtual environment via various display systems, such as head-mounted displays (HMDs), handheld devices, monitors, or optical projection systems. Among these, HMDs are the most prevalent AR devices. An HMD resembles a helmet worn on the forehead, offering a transparent, lightweight, and compact display that can present additional data and information while allowing users to maintain focus on their tasks without significantly obstructing their field of view. Notably, the Microsoft HoloLens is powered by 32 bit architecture processors, incorporating a custom-built Microsoft holographic processing unit, and it benefits from 2 GB of RAM and 64 GB of flash memory. It is equipped with four environment-understanding cameras (including a depth camera), a 2 MP photo/HD video camera, four microphones, and an ambient light sensor. Since its operating system is a simplified version of Microsoft Windows, it can seamlessly integrate with Visual Studio and Unity.

In terms of haptic devices, it is essential to recognize that different HapticAR applications necessitate appropriate haptic feedback provided by specific haptic devices. For instance, ground-based devices (e.g., the Touch haptic device) typically employ force-reflecting joysticks to reconstruct the direction and amplitude of the feedback force. Consequently, these devices are frequently utilized in medical (e.g., IDEA individual dental education assistant [5]) and industrial contexts (e.g., VR4Robots [6]). In contrast, wearable devices (e.g., HaptX gloves) utilize haptic arrays or motors to apply force feedback directly to the skin. Given that wearable devices have a higher number of haptic feedback units compared to ground-based devices, they are better suited for perceiving object properties such as mass, shape, and size. To optimize the haptic feedback force signal rather than focusing on object properties, the Touch haptic device connected to an acupuncture needle is chosen for implementing haptic technology.

Software: The software comprises a haptic generation algorithm, a Unity-based virtual training environment, and a skill assessment subsystem. The virtual environment, developed using unity, is displayed on the Microsoft HoloLens. Within this environment, virtual objects such as the virtual body, the virtual needle, and various user experience components (e.g., menus and buttons) are placed. These virtual objects can be manipulated through two distinct methods. One method is the gesture-based interaction provided by the Microsoft HoloLens, while the other is the touch-based interaction facilitated by the haptic device. Specifically, the stylus of the Touch haptic device is mapped to the virtual acupuncture needle within the virtual environment. Students can control the virtual acupuncture needle to move and insert it into the virtual skin. During the needle insertion process, corresponding virtual changes (e.g., skin deformation) and haptic feedback are rendered both visually and physically. The method of haptic rendering is determined by the relevant generation algorithm, such as the AVHR described in Sect. 5.1 or the approach presented in [7]. For instance, the method in [7] takes the recorded needle motion signals as inputs and reconstructs the haptic signals. Once the needle insertion operation is completed, the skill assessment subsystem is activated. This subsystem calculates the deviations of needle motion signals (position, velocity, acceleration, and force) and the generated haptic signals between the predefined standard manipulation and the online training manipulation, and subsequently produces a reasonable score for the students.

6.3.2 Training Data Collection and Evaluation Metric

Training data collection: To obtain excellent performance, haptic generation algorithms often require massive amounts of data with high-quality labels. In general, a dataset for training should include input data and label data. For input data, the needle motion signals are collected by querying the cache of the Touch haptic device. The motion signals include position p, velocity v, and acceleration a of the needle tip. Note that the values of these signals are relative to the device coordination system. These motion signals will be fed into the haptic generation algorithm for constructing haptic feedback force signals.

For label data, the actual haptic feedback force is hard to measure directly. Thus, we utilize the force-sensing resistor to measure the external force imposed on the needle and calculate the haptic feedback force indirectly under Newton's laws. Specifically, the acupuncture needle is attached to the Touch haptic device by clamps. The force-sensing resistor (FSR) is placed between the clamps and the acupuncture needle (see Fig. 6.5). An Arduino board is used to convert the analog resistance signals from the FSR into the digital force signals. We insert the needle into the skin to let FSR record the external force data. The haptic feedback force is calculated based on Newton's second law, which is given by

$$F_h = F_e - ma, \tag{6.1}$$

where m is the needle mass. a is the needle acceleration. F_h and F_e denote the haptic feedback force and the external force, respectively. Besides, the gravity of the needle is much smaller than F_h and F_e. Therefore, we ignore the gravity in the above equation. In summary, the haptic dataset contains p, v, a, and F_h.

Evaluation metric: As a traditional treatment, acupuncture necessitates precise stimulation of acupuncture points, characterized by accurate position, velocity, acceleration, and force. Moreover, the process of acupuncture needle insertion encompasses multiple phases (e.g., insertion, twisting, and lifting-thrusting phases), which are more complex than those involved in standard needle insertion. To date, a rational

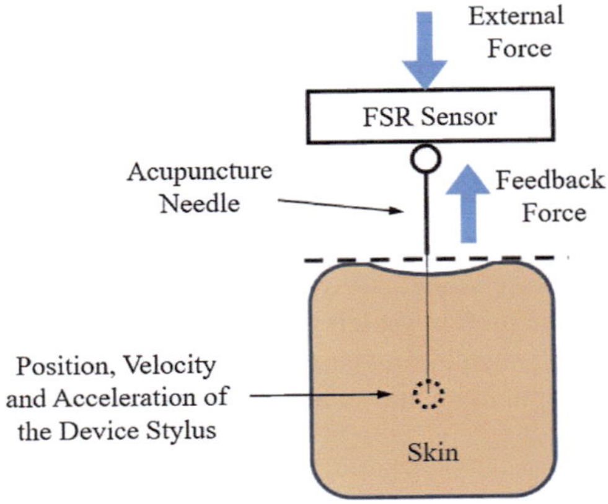

Fig. 6.5 Force effect analysis during data collection process [7]

digital skill assessment framework for acupuncture has not been established. Consequently, an evaluation metric is defined in the following. It quantifies the deviation between the standard signals derived from experts and the actual training signals obtained from students.

$$\begin{aligned} Score = & \lambda_F \left[1 - norm\left(|F_m - F_s|\right)\right] + \lambda_p \left[1 - norm\left(|p_m - p_s|\right)\right] \\ & + \lambda_v \left[1 - norm\left(|v_m - v_s|\right)\right] + \lambda_a \left[1 - norm\left(|a_m - a_s|\right)\right]. \end{aligned} \tag{6.2}$$

Here F_m, p_m, v_m, and a_m represent the recovered haptic feedback force and the recorded position, velocity, and acceleration of the acupuncture needle during online actual training, respectively. Meanwhile, F_s, p_s, v_s, and a_s correspond to the respective values obtained from offline standard operations. The function $norm(\cdot)$ denotes the normalization process that scales the absolute deviation values into the range [0, 1]. Given that different signals contribute differently to acupuncture manipulation, the weighted score is computed, regulated by λ_F, λ_p, λ_v, and λ_a.

6.4 Remote Throat Swab Sampling Platform

During the COVID-19 pandemic, a remote throat swab sampling platform is constructed based on cross-modal communication techniques, which is of great significance in protecting the lives of medical staff and breaking the chains of virus transmission [8]. In other words, the constructed platform, through visual navigation and haptic feedback, is dedicated to remotely controlling a robotic arm to achieve precise throat swab sampling and collection. Specifically, the platform mainly consists of three parts: the master domain, the wireless network, and the slave domain. Specifically, in the master domain, Geomagic Touch (with a sampling rate of 1 kHz) is primarily used to send kinesthetic commands (such as position, velocity, etc.) and receive haptic feedback. In addition, visual information from a laptop can assist the operator in controlling the movement of the robotic arm. Regarding the wireless network, it is mainly responsible for the real-time transmission and flexible scheduling of multi-modal streams. The local network transmission capacity is 30 Mb/s. An edge server (equipped with 8 NVIDIA Tesla M40 GPUs and 12 GB of VRAM) is deployed at an edge node close to the master domain to facilitate the reconstruction of haptic signals. The slave domain mainly comprises three devices: a ZED 2i camera and an AUBO-i3 robotic arm equipped with a magic robotic hand. It is worth noting that the magic robotic hand, with pressure sensors embedded in its five fingers, can measure force feedback within the range of 0 to 10 newtons. To complete remote throat swab sampling, the following three essential technical steps must be included, also shown in Fig. 6.6.

Step 1 (Mouth tracking): When the participant's head is kept stable or fixed, it is first necessary to locate their mouth and then automatically place the cotton swab near the mouth. The key point of this step is to provide the operator with high-definition video services to achieve mouth tracking. Technically, under the guidance of the ZED 2i camera and hand-eye calibration technology, the robotic arm can be navigated to the vicinity of the mouth through the three-dimensional coordinates of the mouth.

Step 2 (Secretion sampling): The cotton swab is manipulated to collect secretions from the throat inside the mouth. Specifically, with the assistance of visual and haptic

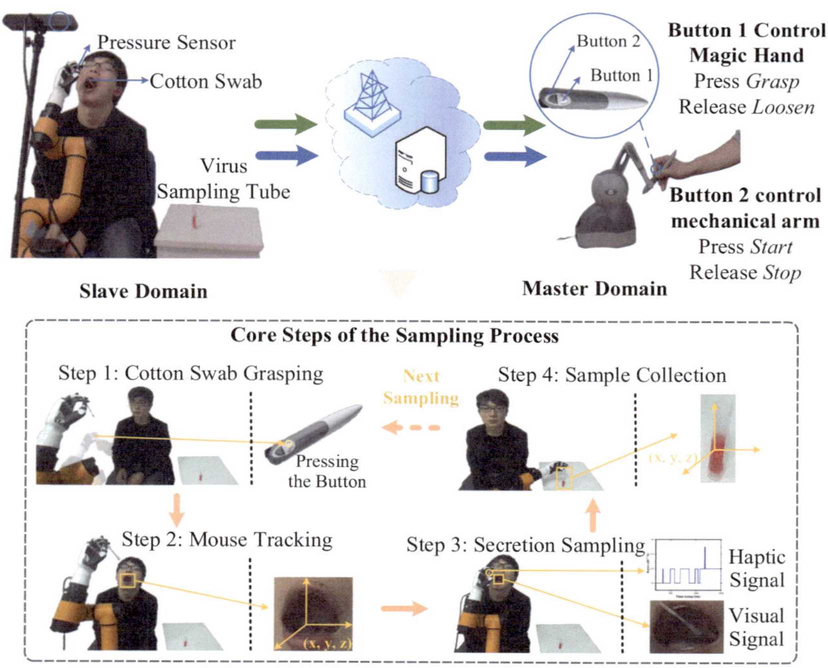

Fig. 6.6 Implementation details and the core steps of throat swab sampling process

feedback, the operator remotely controls the robotic arm. When the operator receives haptic feedback, it indicates that the cotton swab is continuously touching the throat and transmitting pressure feedback to the magic robotic hand. Therefore, the core of this step is how to ensure both haptic fidelity and visual navigation quality to achieve precise and efficient remote sampling.

Step 3 (Sample collection) The cotton swab sample with secretions must be placed into a viral sampling bottle using the robotic arm. Specifically, after completing the throat swab sampling, it is essential to place the sampled swab into the culture fluid through visual navigation. Similar to Step 1, this step can also find the three-dimensional coordinates of the viral sampling bottle through hand-eye calibration technology, and then automatically navigate the cotton swab into the viral sampling bottle.

Figure 6.7 presents the typical haptic signals collected by touching different parts inside the mouth. Specifically, these haptic signals all come from the pressure sensors embedded in the thumb and index finger of the robotic hand, which are mainly responsible for holding the cotton swab. As shown in this figure, when the cotton swab touches the palate, throat, and tongue, respectively, the two pressure sensors receive three different sets of haptic signals. Particularly, the horizontal and vertical axes in this figure represent the time stamp and the magnitude of force, respectively.

Finally, the practicality of the constructed remote throat swab sampling platform is evaluated from three aspects: sampling accuracy, sampling time, and operational experience. Here, a total of 20 participants joined this experiment, with their ages ranging from 23 to 28 years old. From Fig. 6.8, all the 30 participants are able to maintain high remote throat swab sampling accuracy. Through a series of comparative experiments, the average sampling accuracy could reach 97%. Lastly, Table 6.3 also displays the sampling time and operational experience under different sampling platforms. The sampling time refers to the duration from grasping the cotton swab to placing it into the viral sampling bottle. The constructed platform uses a semi-automatic

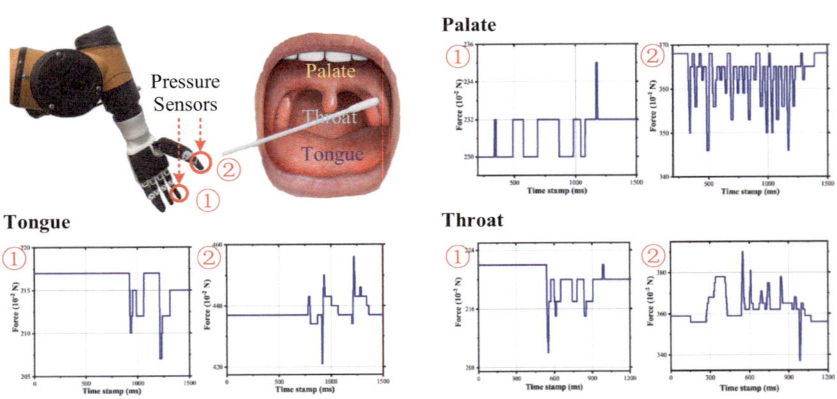

Fig. 6.7 Typical haptic examples collected by the established platform

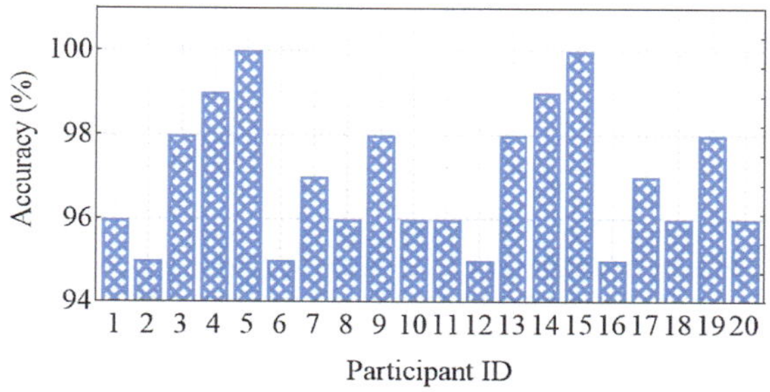

Fig. 6.8 Sampling accuracy of the established platform

Table 6.3 Performance comparison among different platforms

Robot name	Time (s)	Operator's experience
"Lingcai" robot [9]	60~100	Fair
TOOS robot [10]	50~60	Fair
Ours [8]	41±6	Good

control technique. It also has the ability to provide a stable visual stream. Compared with other sampling platforms listed in this table, the constructed platform consumed less sampling time. At the same time, by combining high-quality visual navigation with haptic feedback, operators using this platform could obtain an excellent immersive experience.

6.5 Summary

In this chapter, we describe several cross-modal communication prototype systems. From the descriptions provided above, it can be seen that cross-modal communication techniques can promote the development of multi-modal services in the domains of industry, education, and healthcare. Although the functions and performance of these prototype systems are still very rudimentary and limited, they provide inspiration and valuable experience for the design and implementation of more complicated cross-modal communication systems in the future.

References

1. Li A, Chen J, Wei X, Zhou L (2022) 6G-oriented cross-modal signal reconstruction technology. J Commun 43(6):28–40
2. Li B, Liu X, Dinesh K, Duan Z, Sharma G (2019) Creating a multitrack classical music performance dataset for multimodal music analysis: challenges, insights, and applications. IEEE Trans Multimedi 21(2):522–535

3. Li Y, Zhu J, Tedrake R, Torralba A (2019) Connecting touch and vision via cross-modal prediction. In: Proceedings of the IEEE conference computer vision pattern recognition, Long Beach, USA, June 16–20, pp 10601–10610
4. Sundaram S, Kellnhofer P, Li Y, Zhu J, Torralba A, Matusik W (2019) Learning the signatures of the human grasp using a scalable tactile glove. Nature 569(7758):698–702
5. Roy E, Bakr MM, George R (2017) The need for virtual reality simulators in dental education: a review. Saudi Dental J 29(2):41–47
6. Flaig T, Neugebauer JG, Wapler M (1994) VR4Robots, a new off-line programming system based on virtual reality techniques. In: Proceedings of the international symposium on industrial robots, Hanover, Germany, Apr 25–27, p 671
7. Li A, Chen Y, Ni S, Chen J, Zhou L (2022) Haptic signal reconstruction in eHealth internet of things. IEEE Internet Things 9(18):17047–17057
8. Gao Y, Ni S, Wu D, Zhou L (2022) Edge-based cross-modal communications for remote healthcare. IEEE J Sel Area Comm 40(11):3139–3151
9. Li S, Guo W, Liu H, Wang T, Zhou Y, Yu T, Wang C, Yang Y, Zhong N, Zhang N, Li S (2020) Clinical application of an intelligent oropharyngeal swab robot: implication for the COVID-19 pandemic. Eur Respiratory J 56(2):1–5
10. Chen W, Zhou J, Cheng SS, Lu Y, Zhong F, Gao Y, Wang Y, Xue L, Tong MCF, Liu Y (2021) Tele-operated oropharyngeal swab (TOOS) robot enabled by TSS soft hand for safe and effective sampling. IEEE Trans Med Robot Bionics 3(4):1040–1053

Chapter 7
Extended Techniques

Abstract In this chapter, two extended cross-modal communication techniques are presented. On the one hand, cross-modal communication technology has been deeply integrated with semantic communications, which has attracted widespread attention in recent years, forming a new communication paradigm called cross-modal semantic communications. This new communication paradigm fully explores the semantics both within each modality (intra-modal semantics) and among modalities (inter-modal semantics), and combines them organically. It can effectively solve the polysemy and ambiguity issues existing in current semantic communications and ensure the stability of semantic transmissions. On the other hand, in extremely low communication and computing resource scenarios such as underwater equipment troubleshooting, agricultural operations in remote mountainous areas, and emergency rescue missions responding to natural disasters, existing regular cross-modal communication techniques often fail, and the associated algorithms need to be redesigned. This chapter takes the cross-modal haptic signal recovery in an emergency rescue scenario as an example to illustrate the encountered challenges and available solutions in this direction.

7.1 Cross-Modal Semantic Communications

7.1.1 Necessity and Challenges

From the existing studies, it is evident that the potential correlations between the audio-visual and haptic modalities are crucial when designing communication schemes. These potential correlations can be regarded as "semantic" independent of modality, to be more precise, namely *inter-modal semantics*. By leveraging inter-modal semantics, one modality's content can be effectively compressed from the other modalities when designing cross-modal coding mechanisms, described in Sect. 3. One modality's stream can be efficiently scheduled from the other modalities when constructing cross-modal transmission strategies, described in Sect. 4. One modality's signal can be realistically generated or repaired from other modalities when designing cross-modal signal recovery approaches, described in Sect. 5.

X. Wei et al., *Cross-Modal Communication Technology*, Wireless Networks,
https://doi.org/10.1007/978-3-032-00957-9_7

On the other hand, with the development of 5G, microelectronics, and artificial intelligence, semantic communication has been proposed as a more intelligent communication paradigm [1, 2]. It focuses on the meaning of the transmitted information rather than merely the accurate transmission of bit streams. Compared with bit-based communication, semantic communication has five main characteristics. (1) The source can extract and encode the semantics of the raw message for transmission. The destination should be able to "understand" and infer the message from the received semantics. (2) There are two different types of noise: physical channel noise and semantic noise. Specifically, when extracting and interpreting messages at the source and destination, the semantic noise may seriously affect the performance of semantic communication, as there is ambiguity in words, sentences, or symbols. (3) A semantic communication system is knowledge-based. The source and destination can establish background knowledge bases through self-learning. (4) The performance metrics for measuring semantic communication systems are diverse, depending on the type of original messages or knowledge bases (i.e., text, speech, image, video). (5) Semantic communication can support either full data reconstruction or direct task execution based on the demands of services. Essentially, the "meanings" obtained from respective modality signals, also known as *intra-modal semantics*, are utilized to execute semantic communication.

Generally speaking, semantic communication has several notable advantages. It can effectively compress original messages by filtering out useless, irrelevant, and non-essential information while preserving meaningful information. However, ensuring high reliability remains a significant challenge in semantic communication due to the existence of polysemy and ambiguity issues [3]. After introducing inter-modal semantics in cross-modal communication, the hidden but true meanings can be explored in addition to intra-modal semantics. In other words, semantics in cross-modal communication can effectively address the polysemy and ambiguity problems in semantic communication.

Based on this, the *cross-modal semantic communication*, which combines the respective advantages of cross-modal communication and semantic communication, has emerged [4]. Its basic architecture is shown in Fig. 7.1. Firstly, cross-modal semantic communication focuses on semantics rather than signals (which is fundamentally different from cross-modal communications [5, 6]). By transmitting and processing key information/semantics, the demands for caching, communication, and computation resources are greatly reduced. Secondly, it uses the concept of potential correlation among modalities to guide the perception, transmission, and reception of audio, visual, and haptic-related semantics. While achieving information redundancy reduction, it ensures transmission efficiency and reliability, bringing humans an ultimate interactive experience and rich scenario immersion. However, it is not easy to truly achieve multi-dimensional perception, real-time interaction, and realistic reproduction through cross-modal semantic communication to further promote the development of multi-modal services such as immersive teleoperation. The main difficulties faced are as follows:

Firstly, semantic extraction is challenging. Most existing research focuses on semantic extraction from a single modality (e.g., text, audio, video), and typically

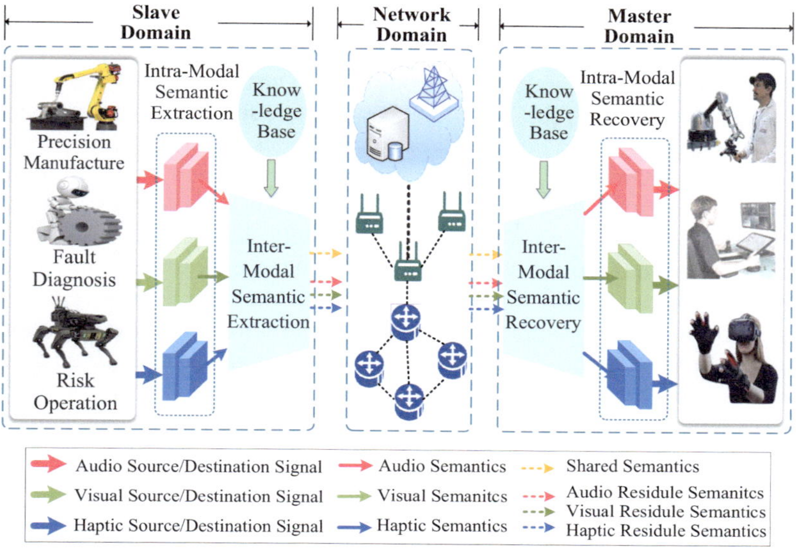

Fig. 7.1 Architecture of cross-modal semantic communications

employs deep learning techniques, which lack interpretability [7]. Even when multiple modalities are involved, the extraction of semantics is usually conducted in an isolated manner, without interference between the semantics of different modalities [8]. In multi-modal services such as immersive teleoperation, robots in the slave domain often have the capability to capture audio, visual, and haptic signals simultaneously, forming a multi-dimensional sensory information source. The collected signals represent a multifaceted view of the scene, objects, or processes they perceive. For each modality signal reflecting sensory information, its semantics alone are insufficient to describe the key information of a complex scene comprehensively, leading to semantic ambiguity. Meanwhile, there is semantic redundancy between multiple sensory-related modality signals. Therefore, at the slave domain, accurately extracting the semantics of each modality and deeply exploring the semantic correlations between modalities to achieve precise perception and understanding of the scene is the primary challenge faced by cross-modal semantic communication research.

Secondly, semantic transmission is difficult. In multi-modal services such as immersive teleoperation, the interaction between the operator at the master domain and the robot at the slave domain requires extremely high real-time performance for transmission. In cross-modal semantic communications, on one hand, the semantics of each modality and the semantics between modalities need to be transmitted, and there are significant differences in their importance. However, existing research often treats the transmission of various types of semantics indiscriminately, making it difficult to ensure transmission reliability [9]. On the other hand, the actual transmission environment is dynamically changing, and the transmission resources and network capacity are very limited. Existing research has not fully considered the impact of the transmission environment and resource consumption on the efficiency and reliability of semantic transmission [10]. Therefore, how to more selectively choose appropriate transmission modes for different types of semantics in a dynamic environment, more efficiently allocate valuable transmission resources such as spectrum, transmit power, and bandwidth, and design semantic transmission control strategies that ensure real-time interaction, will become increasingly challenging.

Thirdly, semantic recovery is challenging. To enhance the realistic reproduction of important content at the master domain in multi-modal services such as immersive teleoperation, the quality of the received semantics needs to be ensured. This is mainly constrained by three factors. (1) The asynchrony of received semantics from different modalities caused by transmission delay directly affects the teleoperation experience. (2) The degradation of received semantic quality due to transmission interference and noise also leads to a decline in the teleoperation experience. (3) The conversion of semantics from each modality to signals for audio-visual and haptic feedback devices determines the final content reproduction effect. Existing researches focuses on using semantic recovery for machine-oriented tasks such as machine translation [11] and multi-modal classification [12], rather than the immersive signal presentation that teleoperation systems are concerned with. Therefore, establishing synchronization of semantics from different modalities, compensating for low-quality semantics, and reconstructing signals for each modality at the master domain to ultimately enhance the immersive teleoperation experience is crucial.

Therefore, to handle these issues, a specific cross-modal semantic communication paradigm for immersive teleoperation application is designed in Sects. 7.1.2–7.1.4. Specifically, it firstly targets the visual, audio, and haptic source signals collected by the robot at the slave domain to achieve semantic extraction within each modality and between modalities. Next, communication delay is a key factor that leads to the decline of interaction stability between the operator and the robot at the master and slave domains. So it is proposed to start with the importance of the semantics to be transmitted. Through measures such as transmission mode selection and transmission resource allocation, the transmission of semantics in each modality can be made adaptable to the dynamically changing environment and limited resources. Finally, the rendering effect of the master domain on the important audio, visual, and haptic content determines the operator's work efficiency. Thus, it is proposed to start from the factors affecting the immersive experience of teleoperation, to achieve synchronization of semantics in different modalities, compensation for low-quality modality semantics, and reconstruction of the destination signals in each modality.

7.1.2 Cross-Modal Semantic Extraction

The designed cross-modal semantic extraction scheme under multi-dimensional perception conditions is as follows: Firstly, the intra-modal semantic information about audio, visual, and haptic modalities is separately extracted from audio, visual, and haptic sensory signals collected by robots in the industrial teleoperation scenario. Secondly, based on probabilistic graphical model theory [13], a graphical model structure that reflects the relationships among the semantics of each modality, the shared semantics, and the residual semantics of each modality is established to achieve the representation of inter-modal semantic information. Finally, the shared semantics and the residual semantics of each modality can be derived.

Intra-modal semantic extraction from each modality source signals

Here, visual, audio, and haptic semantics (denoted as S_V, S_A, S_H) are extracted from visual, audio, and haptic source signals (denoted as V, A, H), respectively. To enhance the interpretability of the extracted semantics from each modality, a combination of handcrafted features and knowledge-based features is proposed (as shown in Fig. 7.2), which is detailed as follows:

For the *semantic extraction of visual source signals*, the process is divided into four steps: key-frame localization, region segmentation, object recognition, and semantic parsing.

(1) Key-frame localization in the videos is achieved using the frame-difference method. Specifically, the first frame of the video is designated as the key frame (benchmark), and a certain threshold is set. When the difference between a subsequent frame and the first frame exceeds this threshold, the subsequent frame is considered a key frame. This frame then becomes the new benchmark to continue searching for subsequent key frames. Through this iterative process, all key frames in the video are obtained.
(2) For each key frame, region growing is employed. Starting from an initial region, adjacent pixels or other regions with the same properties are merged into the current region until the region exceeds a certain threshold. This method completes the region segmentation of the current visual key frame.
(3) The visual knowledge base is assumed to contain images of typical objects. It also includes a CNN that describes the mapping relationship between visual features and typical object features. This deep network is used to match the visual features of each key frame and each segmented region in the video with the object features in the knowledge base. This process identifies the objects $\{o_1^{(V)}, \ldots, o_n^{(V)}, \ldots, o_N^{(V)}\}$ present in each key frame and each segmented region of the video.
(4) For the identified objects, their semantics are parsed from three perspectives: temporal, spatial, and static visual attributes. Temporally, the tracking method such as the Kalman filter is used to obtain the motion and change trajectory $T_n^{(V)}$ of the current object $o_n^{(V)}$. Spatially, a relational template matching method is proposed to determine the position and relative relationships $R_n^{(V)}$ between $o_n^{(V)}$ and the background of the current frame, as well as other objects. In terms of static visual attributes, the color, shape, and texture attributes $A_n^{(V)}$ of $o_n^{(V)}$ are computed using corresponding feature operators, such as extracting color based

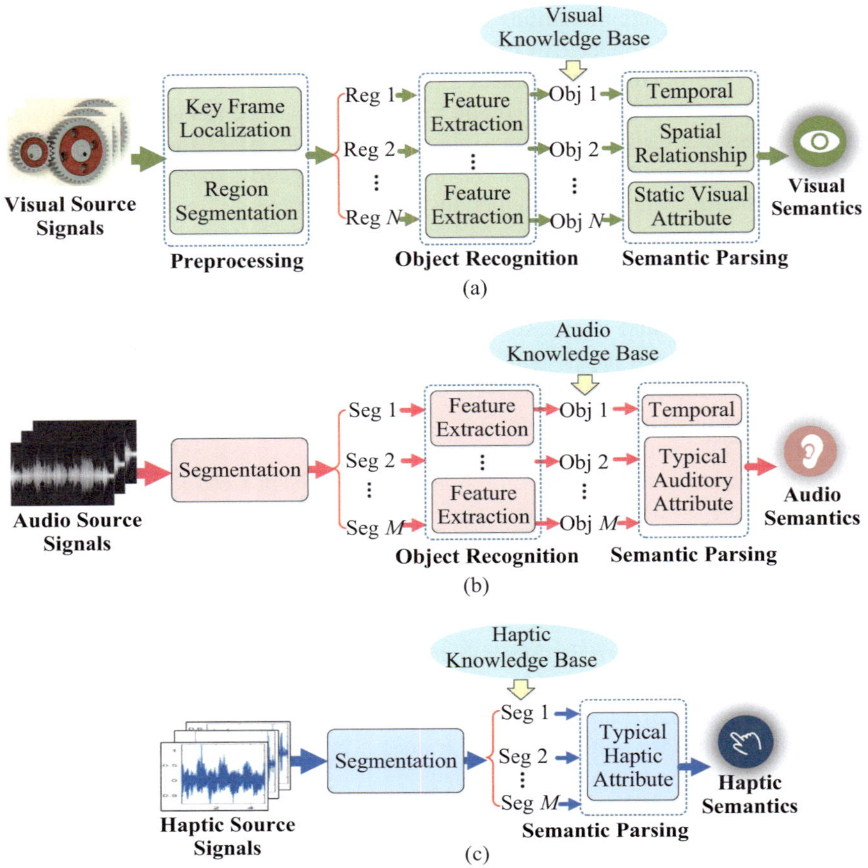

Fig. 7.2 Illustration of intra-modal semantic extraction: **a** Visual, **b** Audio, **c** Haptic

on color histograms, shape based on edge detection operators, and texture based on local binary patterns.

Through the aforementioned methods, the transformation from the visual source signal V to visual semantics S_V is achieved. Specifically, with visual key frames and the objects within each key frame as the center, the motion, spatial relationships, and static visual attributes (color/shape/texture) of each object are parsed and collectively represented as the visual semantics $S_V = \{T_n^{(V)}, R_n^{(V)}, A_n^{(V)}\}_{n=1,\ldots N}$.

For *the semantic extraction of audio source signals*, considering that audio signals are one-dimensional and exhibit characteristics such as short-term stationarity and long-term variability, the process is divided into three steps: key-segment segmentation, object recognition, and semantic parsing.

(1) The audio signal is segmented into a series of meaningful audio segments using a method that combines energy thresholding with typical audio events.
(2) Each audio segment is input into a pre-trained RNN in the audio knowledge base, which reflects the mapping relationship between audio segment features and typical object features. Feature matching with the typical audio in the knowledge base identifies the objects $\{o_1^{(A)}, \ldots, o_m^{(A)}, \ldots, o_M^{(A)}\}$ within each segment.
(3) For the identified objects, their semantics are characterized from two perspectives: temporal and typical auditory attributes. In the temporal dimension, the duration $T_m^{(A)}$ of the current object $o_m^{(A)}$ is extracted. In terms of typical auditory attributes, the frequency and amplitude of $o_m^{(A)}$ are obtained from the spectrogram of the audio signal. The timbre is determined by the distribution of high frequencies and their respective amplitudes in the spectrum, collectively represented as $A_m^{(A)}$.

Through the aforementioned methods, the transformation from the audio source signal A to audio semantics S_A is achieved. Specifically, with the objects within each audio segment as the center, the duration, frequency/amplitude/timbre of each object are parsed and collectively represented as the audio semantics $S_A = \{T_m^{(A)}, A_m^{(A)}\}_{m=1,\ldots M}$.

For *the semantic extraction of haptic source signals*, as haptic signals contain the least amount of information compared to audio and visual signals, the typical haptic attributes from each haptic segment are directly used as the semantics S_H. It is worth noting that [14] provides 15 representative haptic features. Among them, for the most common object surface haptic signals, the five most important features are: friction, texture (smoothness/roughness), thermal conductivity, compliance, and adhesion. Through the aforementioned methods, the transformation from the haptic source signal H to haptic semantics S_H is realized.

Inter-modal semantic construction by probabilistic graphical model

After extracting the semantics of visual, audio, and haptic source signals, we start from the fact that the semantics of different modalities provide a multifaceted description of the scene, perceived objects, and executed processes by the robot at the slave domain. With the intra-semantics of visual, audio, and haptic modalities as explicit information, we define the *shared semantic* and the *residual semantic of each modality*, treating them as implicit elements. We then establish the relationship between explicit information and implicit elements. On one hand, each modality's semantics are represented as interpretable shared semantics and residual semantics of each modality. On the other hand, under the combined influence of compression requirements and knowledge rules, we aim to suppress semantic ambiguity while obtaining the minimum semantic bitstream.

Specifically, the shared semantics refer to the semantics that are not dependent on specific modalities and have intrinsic consistency. In other words, they represent the correlation between different modalities. For example, for a certain object, it has a circular appearance with teeth (visual semantic), makes a unique sound when rotat-

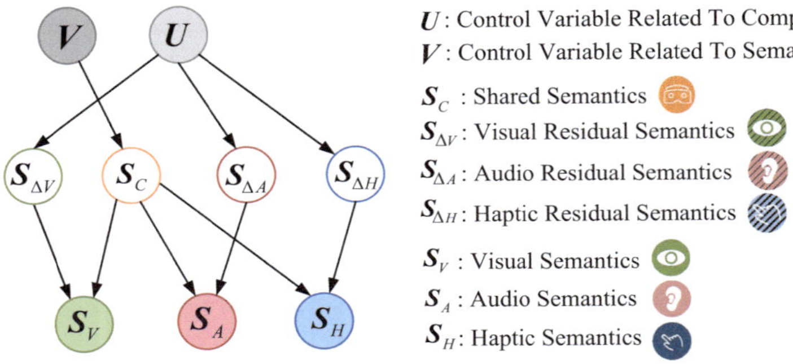

Fig. 7.3 Illustration of inter-modal semantic construction

ing (audio semantic), and is made of a hard material (haptic semantic). Its shared semantics would be "gear". Residual semantics of each modality, on the other hand, refer to the remaining semantics of each modality after removing the shared semantics, reflecting the characteristics of each modality itself. For example, part of the object's surface is red (visual residual semantic), the sound has a high frequency (audio residual semantic), and it has a rough touch (haptic residual semantic).

Based on the above analysis and probabilistic graphical model theory, we establish the relationship between the semantics of visual, audio, and haptic modalities (explicit information), shared semantics and residual semantics of each modality (implicit elements), compression control variables, and rule control variables. The structure of this probabilistic graphical model is shown in Fig. 7.3, with the specific process as follows:

(1) The model has a hierarchical structure of "semantics of visual, audio, and haptic modalities", "shared semantics & residual semantics of each modality", and "control variables". The bottom layer consists of the extracted semantics of each modality S_V, S_A, S_H, which belong to explicit information. The middle layer includes shared semantics (denoted as S_C) and residual semantics of each modality (denoted as $S_{\Delta V}$, $S_{\Delta A}$, $S_{\Delta H}$, which are implicit elements to be calculated and inferred. The top layer consists of the set of control variables (denoted as $f U, V$).
(2) The relationship between the semantics of each modality and the shared semantics, as well as the residual semantics, is established. For S_V, according to the theory of probabilistic graphical models, it can be regarded as generated jointly by S_C and $S_{\Delta V}$, that is, $S_V \sim f_V(S_C, S_{\Delta V})$. Similarly, for audio and haptic modalities, we have $S_A \sim f_A(S_C, S_{\Delta A})$, $S_H \sim f_V(S_C, S_{\Delta H})$. Here, $f_V(\cdot)$, $f_A(\cdot)$, $f_H(\cdot)$ can be selected as explicit probability distribution functions or described using deep neural networks.
(3) The relationship between the shared semantics and the residual semantics of each modality with control variables should also be established, $\boldsymbol{S}_{\Delta V}, \boldsymbol{S}_{\Delta A}, \boldsymbol{S}_{\Delta H} \sim \bar{g}(\boldsymbol{U})$, $\boldsymbol{S}_C \sim g(\boldsymbol{V})$. The selection principles for $g(\cdot)$ and $\bar{g}(\cdot)$ are similar to those

for $f_V(\cdot)$, $f_A(\cdot)$, $f_H(\cdot)$. It should be noted that U is the control variable related to compression, which affects the residual semantics of each modality according to the current teleoperation task and compression target. V is the control variable related to the predefined semantic association rules about each modality's knowledge base. They are used to guide the extraction of shared semantics from each modality's semantics to enhance the interpretability.

Shared and residual semantic encoding

After the model construction is completed, the shared semantics and residual semantics of each modality (implicit elements) are derived and encoded based on the semantics of each modality (explicit information) and under the influence of the control variable. Specifically, an alternating update mechanism for the shared semantics and the residual semantics of each modality is proposed.

(1) For the shared semantics, it can be derived as

$$S_C \sim \varphi\left(S_V, S_A, S_H, \tilde{S}_{\Delta V}, \tilde{S}_{\Delta A}, \tilde{S}_{\Delta H}, V\right). \tag{7.1}$$

That is, S_C is jointly determined by the semantics of each modality S_V, S_A, S_H, the control variable V, and the residual semantics $\tilde{S}_{\Delta V}$, $\tilde{S}_{\Delta A}$, $\tilde{S}_{\Delta H}$ obtained from the previous iteration.

(2) For the residual semantics, they can be derived as

$$S_{\Delta V} \sim \phi_V\left(S_V, \tilde{S}_C, U\right), \quad S_{\Delta A} \sim \phi_A\left(S_A, \tilde{S}_C, U\right), \quad S_{\Delta H} \sim \phi_H\left(S_H, \tilde{S}_C, U\right). \tag{7.2}$$

That is, $S_{\Delta V}$ is determined by S_V, U and $\tilde{S}_C$ obtained from the previous iteration. $S_{\Delta A}$ is determined by S_A, U and $\tilde{S}_C$. $S_{\Delta H}$ is determined by S_H, U and $f\tilde{S}_C$.

(3) To guide the encoding of shared semantics and residual semantics of each modality, it is necessary to design corresponding objective functions. Two main factors can be considered. On one hand, for visual modality, S_C and $S_{\Delta V}$ should be as close to S_V as possible, i.e., by maximizing their mutual information. The audio and haptic modalities are similar to the visual modality. On the other hand, the control variables U and V are used to adjust S_C and $S_{\Delta V}$, $S_{\Delta A}$, $S_{\Delta H}$. Specifically, the predefined semantic association rules guide the encoding fS_C, and the overall compression rate guides the encoding of $S_{\Delta V}$, $S_{\Delta A}$, $S_{\Delta H}$.

Under the guidance of the objective function, according to the probabilistic graphical model, the derivation and encoding of shared semantics and residual semantics of each modality in Eqs. (7.1) and (7.2) can be implemented using Markov Chain Monte Carlo methods.

7.1.3 Cross-Modal Semantic Transmission

Taking into account the importance of the semantics to be transmitted, the dynamically changing transmission environment, and the constrained transmission resources, cross-modal semantic transmission is realized through the optimized allocation of transmission modes and resources. Specifically, the significance of shared semantics and residual semantics of each modality is firstly analyzed, and the associated transmission modes matching them are designed. Then, a relationship model is established between transmission resource overheads (e.g., spectrum, transmission power, bandwidth) and transmission metrics that affect real-time interaction (e.g., delay, packet loss, bit rate). Finally, aiming to ensure the real-time interaction between the master and slave domains and in response to the dynamically changing transmission environment, an adaptive transmission control strategy based on broad reinforcement learning (BRL) is designed.

Transmission mode selection based on semantic importance

In cross-modal semantic communications, the shared semantics, visual/audio/haptic residual semantics coexist after extraction. However, for immersive teleoperation, the transmission requirements and importance of these different types of semantics vary significantly. Specifically, since the shared semantics are derived through compression and refinement based on the semantics of each modality, they have relatively small capacity. Yet, they are the most important information for the master domain. Therefore, it is essential to ensure that the transmission of the shared semantics is as error-free as possible and has the highest level of security.

The significance of visual/audio/haptic residual semantics is clearly weaker than that of shared semantics and depends on the specific teleoperation task. Overall, haptic residual semantics have stringent requirements for transmission delay and reliability [15]. For audio-visual residual semantics, their relatively large capacity demands more available spectrum and bandwidth [16]. Given this, a combination of multiple transmission modes is proposed, shown in Fig. 7.4.

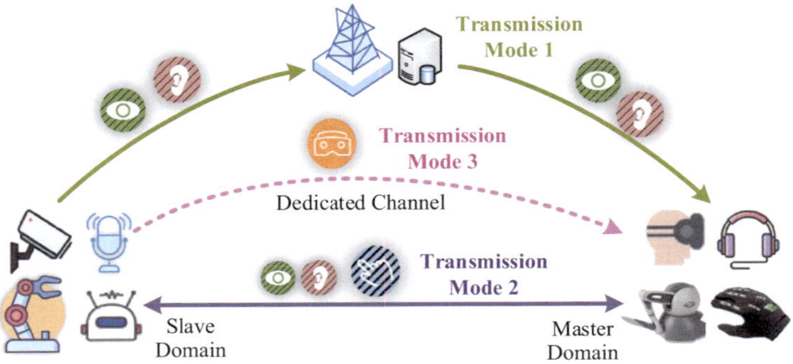

Fig. 7.4 Transmission modes for the shared and the residual semantics

Mode 1: The robot at the slave domain and the feedback device at the master domain complete semantic transmission through the base station. That is, the semantics from the slave terminal are transmitted to the base station via a common service channel, the base station processes and transmits the semantics to the master domain.

Mode 2: A direct communication method is used between the master and slave domains. That is, terminals at the slave domain directly transmit the semantics to the master domain via a common service channel without going through the base station, and the reception and processing of the semantics are directly completed at the master domain. At the same time, terminals at the master domain can also use this method to send control commands to the slave domain.

Mode 3: A dedicated channel is used for semantic transmission. A dedicated channel is an independent channel allocated for a specific purpose, with the main characteristics of exclusivity and security. This dedicated channel can be supported by the base station or relevant devices in direct communication.

Based on the above transmission modes, the related transmission rules can be designed.

(1) For the shared semantics, mode 3 is adopted. The reason is that the high communication quality and security provided by the dedicated channel can effectively avoid interference from other transmission paths during transmission of the shared semantics.
(2) For the haptic residual semantics, mode 2 is selected. The purpose is twofold. On one hand, the haptic bitstream size is relatively small, and the required bandwidth for direct communication is small, ensuring low latency. Direct reception and processing of haptic semantics at the master domain will not consume too many caching, communication, and computation resources. On the other hand, in immersive teleoperation, haptic modality plays a role in the control and feedback interaction between the master and slave domains. Using direct communication can ensure fast and reliable two-way interactions.
(3) For the audio and visual residual semantics, a hybrid transmission method mainly based on mode 1 and supplemented by mode 2 is used. The purpose is that the scale of visual and audio residual semantics is relatively large. With the strong transmission capability and bandwidth guarantee of the base station, fast and smooth transmission can be achieved, so mode 1 is preferred. When mode 2 is used, the priority of transmitting audio and visual residual semantics and haptic residual semantics is further determined according to the type of teleoperation task: for teleoperation tasks mainly focused on audio and visual perception and feedback, audio and visual residual semantics are prioritized for transmission. For tasks mainly focused on haptic perception and feedback, as well as tasks where audio, visual, and haptic perception and feedback are equally important, haptic residual semantics are prioritized for transmission.

Collaborative resource allocation model construction

Based on the selected transmission mode, the associated collaborative resource allocation modeling scheme should be designed. Its aim is to achieve the collaborative allocation of spectrum, transmission power, and bandwidth resources, and focusing on key indicators that affect real-time interactive performance, such as transmission delay, packet loss, and bit rate.

(1) For the shared semantics, the transmission delay D_C includes transmission delay, propagation delay, and reception delay. Given a fixed total amount of shared semantic transmission, D_C is related to the bandwidth B_3 of the dedicated channel, that is, $D_C \sim \gamma_3(B_3)$. Additionally, the packet loss L_C is related to the allocated spectrum F_3 and transmission power P_3, that is, $L_C \sim \xi_3(F_3, P_3)$. The bit rate R_C is closely related to the bandwidth B_3 and transmission power P_3, that is, $R_C \sim \eta_3(B_3, P_3)$.

(2) For the haptic residual semantics, the transmission delay $D_{\Delta H}$ includes transmission delay, propagation delay, reception delay, and queuing delay when competing with audio and visual residual semantics for access to the direct communication channel, $D_{\Delta H} \sim \gamma_2(B_2)$. Similarly, for packet loss and bit rate, we have $L_{\Delta H} \sim \xi_2(F_2, P_2)$, $R_{\Delta H} \sim \eta_2(B_2, P_2)$.

(3) For the audio and visual residual semantics, when using mode 2, the transmission delay, packet loss, and bit rate are similar to those of haptic residual semantics. However, when using mode 1, the transmission delay includes transmission delay, propagation delay, reception delay, and the queuing delay of semantic data packets waiting for processing at the base station. Packet loss is related to the spectrum and transmission power of the uplink and downlink. The bit rate is also related to the bandwidth and transmission power of the uplink and downlink. Considering both cases, we obtain

$$
\begin{aligned}
D_{\Delta A}, D_{\Delta V} &\sim \alpha \times \gamma_1(B_1) + (1-\alpha) \times \gamma_2(B_2), \\
L_{\Delta A}, L_{\Delta V} &\sim \alpha \times \xi_1(F_1, P_1) + (1-\alpha) \times \xi_2(F_2, P_2), \\
R_{\Delta A}, R_{\Delta V} &\sim \alpha \times \eta_1(B_1, P_1) + (1-\alpha) \times \eta_2(B_2, P_2),
\end{aligned} \tag{7.3}
$$

where the ratio $\alpha \in [0, 1]$.

(4) Based on the above steps, the overall delay $D = \{D_C, D_{\Delta H}, D_{\Delta A}, D_{\Delta V}\}$, packet loss $L = \{L_C, L_{\Delta H}, L_{\Delta A}, L_{\Delta V}\}$, and data rate $R = \{R_C, R_{\Delta H}, R_{\Delta A}, R_{\Delta V}\}$ for the shared semantics and the residual semantics of each modality can be obtained. Additionally, since spectrum, bandwidth, and power resources are very valuable in practical scenarios, it is necessary to consider and meet the corresponding constraints, which are denoted by $\Omega(B, F, P)$. It should be noted that here B represents the set $\{B_1, B_2, B_3\}$, F represents the set $\{F_1, F_2, F_3\}$, and P represents the set $\{P_1, P_2, P_3\}$.

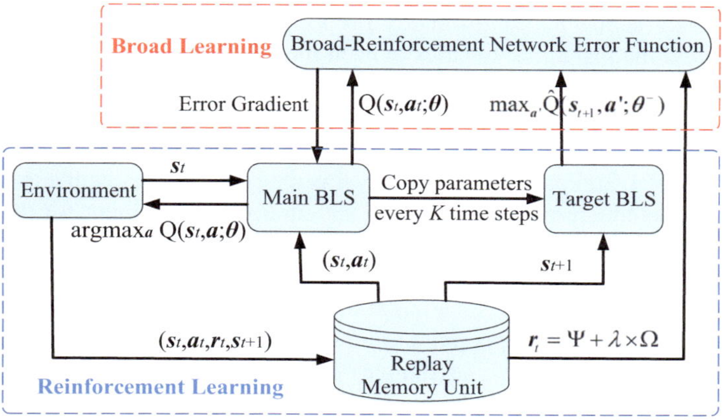

Fig. 7.5 Schematic diagram of adaptive transmission control based on BRL

Design of adaptive transmission control scheme

During the cross-modal semantic transmission process, the transmission environment dynamically changes with the wireless link conditions and the mobility of the master and slave domains. These transmission environment conditions are usually unknown in advance. Therefore, it is necessary to design an adaptive transmission control scheme for dynamic environments to ensure real-time interaction. Given this, broad reinforcement learning (BRL) [17], which combines broad learning system (BLS) with reinforcement learning architecture, can be utilized here, shown in Fig. 7.5.

Specifically, the selected agent is the edge server responsible for the transmission control of the shared semantics and residual semantics of each modality between the master and slave domains in its area. With the goal of ensuring interaction quality and reducing transmission resource overhead, the agent uses BRL to adaptively regulate spectrum, transmission power, and bandwidth resources in the semantic transmission process according to the dynamically changing environment. Here, the transmission delay, packet loss, and bit rate of the terminal are defined as the *state*. That is, the state s_t at time t is $\{D(t), L(t), R(t)\}$. The spectrum, bandwidth, and transmission power are defined as the *action*. That is, the action a_t at time t is $\{B(t), F(t), P(t)\}$. In addition, the *reward* function r_t is defined as the comprehensive benefit composed of real-time interaction quality and transmission resource overhead in the time interval $(t-1, t]$,

$$r_t = \psi(D(t), L(t), R(t)) + \lambda \times \Omega(B(t), F(t), P(t)), \tag{7.4}$$

where $\psi(\cdot)$ is a mapping function reflecting real-time interaction quality, which is related to transmission delay, packet loss, and bit rate. To ensure real-time and reliable transmission, $\psi(\cdot)$ is inversely proportional to the transmission delay and packet loss of the shared semantics and the residual semantics of each modality (especially the haptic residual semantics) and directly proportional to the bit rate of the shared semantics and the residual semantics of each modality (especially audio and visual residual semantics). The parameter λ is used to control the weight of $\psi(\cdot)$ and $\Omega(\cdot)$.

After determining s_t, a_t, and r_t, the BRL can be constructed, containing four main components: environment, replay memory unit, evaluation BLS (E-BLS), and target BLS (E-BLS). For the *environment*, it is the territory the agent engages with. The agent takes the current state s_t from the environment as input for the E-BLS, and then produces an action a_t. Moreover, the subsequent state s_{t+1} and reward r_t are also updated from the environment. For the *replay memory unit*, it caches the experience of the agent (s_t, a_t, s_{t+1}, r_t) at time t. Moreover, it serves as a repository and processing hub for the upcoming training sessions of the E-BLS. It specifically picks out and gathers a batch of samples from the experience base in a random manner, represented as $\{(s_t, a_t, s_{t+1}, r_t)\}_{\mathcal{P}}$. It also prepares the training samples by forwarding $\{s_{t+1}\}_{\mathcal{P}}$ to the T-BLS, obtaining the feedback on the estimated optimal action-values and determining the target values $\{y_t\}_{\mathcal{P}}$. For the *evaluation BLS*, it employs the BLS to establish the relationship between the pairs $\{(s_t, a_t)\}_{\mathcal{P}}$ and their corresponding targets $\{y_t\}_{\mathcal{P}}$. To elaborate, the action-value function $Q(s, a; \boldsymbol{\theta})$ is represented by the E-BLS, where $\boldsymbol{\theta}$ represents the parameters that are optimized to reduce the discrepancy between y_t and $Q(s_t, a_t; \theta)$. Additionally, upon receiving the current state from the environment, it computes Q and determines the best action for the current time step once the training is complete. It also periodically transmits the newly updated parameters to the T-BLS. For the *target BLS*, it is also a BLS. Its structure $\hat{Q}$ and parameters $\boldsymbol{\theta}^-$ are synchronized with E-BLS. Its role is to provide optimal action values after obtaining inputs from the training pool. Based on this, the specific realization of BRL can be derived, which can be divided into six main steps.

Step 1 (Initialization): The structure of the E-BLS are randomly set, containing the number of mapped features, enhancement nodes, and the associated parameters $\boldsymbol{\theta} = \left\{W_n^m, W_f, \boldsymbol{\delta}_f, W_g, \boldsymbol{\delta}_g\right\}$ [18]. The initialization of T-BLS is the same as that of E-BLS. The capacities of the experience base and the repository in the replay memory unit, respectively denoted as $\mathcal{K}$ and $\mathcal{P}$, should also be set.

Step 2 (Action decision-making): When the E-BLS finishes initialization or receives s_t from the environment, it computes the associated action-value function Q. By considering Q, the current a_t can be derived. Here, the ε-greedy policy is adopted. It chooses a random a_t with probability ε, otherwise, it chooses

$$a_t = \operatorname{argmax}_a Q(s_t, a; \boldsymbol{\theta}). \tag{7.5}$$

Step 3 (State and reward caching): The agent implements a_t, enters the next state s_{t+1}, and gets r_t by interacting with the environment. Besides, it caches transition (s_t, a_t, s_{t+1}, r_t) in $\mathcal{K}$. When the number of transitions exceeds the volume of $\mathcal{K}$, the new entering transition replaces the oldest one.

Step 4 (Training sample preparation): To guarantee precise decision-making, the structure and parameters of the E-BLS need to be trained in time. In the E-BLS, simple ridge regression is used for finding the relationship between (s_t, a_t) and its target. Firstly, batch of transitions are randomly pumped from $\mathcal{K}$ and put into $\mathcal{P}$ as training samples, forming $\{(s_t, a_t, s_{t+1}, r_t)\}_{\mathcal{P}}$. Then, $\{s_{t+1}\}_{\mathcal{P}}$ in $\mathcal{P}$ are fed into the T-BLS, and $\left\{\max_{a'} \hat{Q}\left(s_{t+1}, a'; \boldsymbol{\theta}^-\right)\right\}_{\mathcal{P}}$ are obtained. Subsequently, targets $\{y_t\}_{\mathcal{P}}$ are computed in $\mathcal{P}$ by concerning $\{r_t\}_{\mathcal{P}}$ and $\left\{\max_{a'} \hat{Q}\left(s_{t+1}, a'; \boldsymbol{\theta}^-\right)\right\}_{\mathcal{P}}$,

$$y_t = \begin{cases} r_t, & s_{t+1} \text{ is a terminal state} \\ r_t + \gamma \max_{a'} \hat{Q}\left(s_{t+1}, a'; \boldsymbol{\theta}^-\right), & \text{otherwise.} \end{cases} \tag{7.6}$$

Finally, once the accumulated samples and their associated targets have amassed to a sufficient volume, the dataset $\{(s_t, a_t, y_t)\}_{\mathcal{P}}$ becomes available for the training of the E-BLS. $\mathcal{P}$ is emptied, and the process of preparing a new batch of training samples can commence.

Step 5 (E-BLS training): $\{(s_t, a_t)\}_{\mathcal{P}}$ is taken as input X, while $\{y_t\}_{\mathcal{P}}$ from Eq. (7.6) is found as output Y. Firstly, X is mapped to the feature space by using

$$Z_i = \varphi\left(XW_{f_i} + \boldsymbol{\delta}_{f_i}\right), \quad i = 1, 2, \ldots, n, \tag{7.7}$$

where W_{f_i} and $\boldsymbol{\delta}_{f_i}$ are randomly generated. $\varphi(\cdot)$ is a nonlinear mapping such as a logistic function. Then, the mapped features are further processed to obtain enhancement nodes,

$$G_j = \xi\left([Z_1, \ldots, Z_n] W_{g_j} + \boldsymbol{\delta}_{g_j}\right), \quad j = 1, 2, \ldots, m, \tag{7.8}$$

where W_{g_j} and $\boldsymbol{\delta}_{g_j}$ are randomly generated. $\xi(\cdot)$ is also a nonlinear mapping function. Finally, the concatenation of mapped features and enhancement nodes is imported into the output layer, calculated as

$$Y = [Z_1, \ldots, Z_n \mid G_1, \ldots, G_m] W_n^m. \tag{7.9}$$

Here, the connecting weights W_n^m is estimated by ridge regression. Specifically, let $J_n^m = [Z_1, \ldots, Z_n \mid G_1, \ldots, G_m]$. W_n^m can be obtained by pseudoinverse of A_n^m

$$W_n^m = \left(J_n^m\right)^+ \cdot Y = \left(J_n^m \cdot \left(J_n^m\right)^T + \lambda I\right)^{-1} \cdot \left(A_n^m\right)^{\mathrm{T}} \cdot Y \tag{7.10}$$

Is the trained BLS does not achieve the desired performance metrics, such as accuracy, one approach to address this is to augment the enhancement nodes or/and features over several rounds of iteration. The detail of this can be found in [18]. Compared with deep neural network-related models, the E-BLS is capable not only of autonomously identifying an appropriate structure but also of updating parameters that are exclusively linked to newly integrated components. This capability can effectively ensure the autonomous decision-making capacity of BRL while also reducing both training and response times. After training, $\boldsymbol{\theta} = \left\{W_n^m, W_f, \boldsymbol{\delta}_f, W_g, \boldsymbol{\delta}_h\right\}$ can be estimated and $Q\,(s, a; \boldsymbol{\theta})$ is also updated.

Step 6 (T-BLS synchronization): Every K times of E-BLS training, it is cloned to obtain the T-BLS, finishing synchronization. That is, $\hat{Q} = Q, \boldsymbol{\theta}^- = \boldsymbol{\theta}$. Here, $\hat{Q}\left(s, a; \boldsymbol{\theta}^-\right)$ will be used to generate $\max_{a'} \hat{Q}\left(s_{t+1}, a'; \boldsymbol{\theta}^-\right)$ for the following updates to y_t.

By using the above method, it can automatically track environmental changes, adaptively adjust the transmission mode of shared semantics and residual semantics of each modality, and make self-organized decisions on the consumed transmission resources. This approach can improve the efficiency of semantic transmission and provide transmission guarantees for real-time interaction.

7.1.4 Cross-Modal Semantic Recovery

When considering cross-modal semantic recovery, it can be further divided into three components based on the key influencing factors. Firstly, adhering to self-supervised learning, and under the constraint of synchronous transmission and reception of semantics across visual, audio, and haptic modalities, the initial semantics of each modality are decoded from the received shared semantics and residual semantics of each modality. Secondly, leveraging the interrelationships among the semantics of different modalities and employing mutual information maximization and adversarial learning, the low-quality modality semantics can be compensated. Finally, knowledge bases for each modality are constructed, and the destination signals of each modality are reconstructed through semantic-signal mapping, thereby achieving a realistic rendering of important content.

Semantic decoding under modality synchronization constraints

Due to inevitable interference and noise during semantic transmission, there is a difference between the semantics at the slave and master domains. For rigor, the received shared semantics, visual residual semantics, audio residual semantics, and haptic residual semantics are denoted as $\overline{S}_C, \overline{S}_{\Delta V}, \overline{S}_{\Delta A}, \overline{S}_{\Delta H}$.

Different from single-modality semantic decoding, the greatest challenge here is that there is no prior information available to guide the decoding process. Therefore, it is necessary to fully explore the intrinsic relationships within each modality and among modalities. It should be emphasized that semantic synchronization among visual, audio, and haptic modalities is crucial for an immersive experience. Thus, modality synchronization is a core element that needs to be considered in semantic decoding. The specific process is as follows (as shown in Fig. 7.6):

(1) For visual modality, $\overline{S}_C$ $\overline{S}_{\Delta V}$ are used as inputs to construct a visual semantic decoding model to obtain the initial visual semantics $\overline{S}_V$. Here, the relationship between the inputs and the output can be expressed as $\overline{S}_V \sim h_V(\overline{S}_C, \overline{S}_{\Delta V})$, where

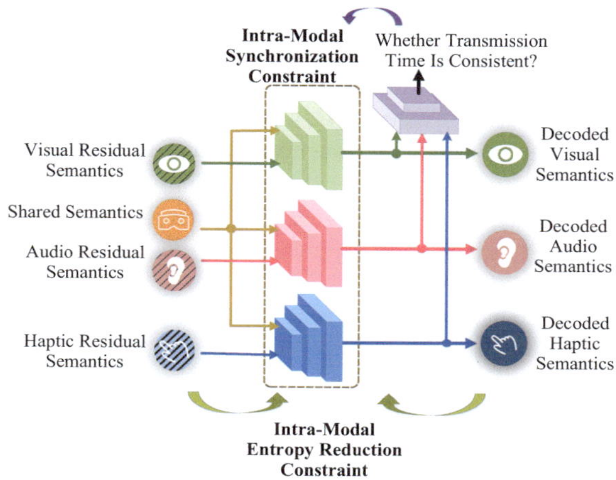

Fig. 7.6 Illustration of semantic decoding under modality synchronization constraints

the semantic decoding model $h_V(\cdot)$ can be based on a deep neural networks or a traditional machine learning model.

(2) The intrinsic information of the visual modality is utilized to guide semantic decoding. The uncertainty should be significantly reduced after semantic decoding. According to semantic information theory [19], entropy is a measure of the orderliness of information. Therefore, the entropy of the decoded visual semantics $H(\overline{\mathbf{S}}_V)$ should be lower than the combined entropy of the shared semantics and the residual visual semantics before decoding $H(\overline{S}_C, \overline{S}_{\Delta V})$. Based on this, the objective function for visual modality semantic decoding can be defined as

$$decode_V = H\left(\overline{s}_C, \overline{s}_{\Delta V}\right) - H\left(\overline{S}_V\right). \tag{7.11}$$

By maximizing $decode_V$ to optimize $h_V(\cdot)$, semantic decoding of visual modality can be achieved.

For the audio and haptic modalities, the same approach as the visual modality is adopted, i.e., the relationships between the inputs and the output before and after decoding are expressed as $\overline{S}_A \sim h_A\left(\overline{S}_C, \overline{S}_{\Delta A}\right)$, $\overline{S}_H \sim h_H\left(\overline{S}_C, \overline{S}_{\Delta H}\right)$, and the corresponding objective functions for audio and haptic modality semantic decoding are

$$\begin{aligned} decode_A &= H\left(\overline{S}_C, \overline{S}_{\Delta A}\right) - H\left(\overline{S}_A\right), \\ decode_H &= H\left(\overline{S}_C, \overline{S}_{\Delta H}\right) - H\left(\overline{S}_H\right). \end{aligned} \tag{7.12}$$

By maximizing the above objective functions to optimize $h_A(\cdot), h_H(\cdot)$, semantic decoding of the audio and haptic modalities can be achieved.

(3) Based on the semantic decoding within each modality, it is necessary to maintain synchronization among the decoded semantics of each modality. Here, the

obtained $\overline{S}_V, \overline{S}_A, \overline{S}_H$ are input into a discriminator and the idea of self-supervised learning is adopted [20]. The purpose of this discriminator is to determine whether the transmission times of the decoded semantics about three modalities are consistent, *i.e.*, whether the three modalities satisfy the synchronization condition. It should be noted that the transmission time can be written into relevant fields when the shared semantics and residual semantics of visual, audio, and haptic modalities are sent. Therefore, the synchronization constraint among modalities can be expressed as $Syn\left(\overline{S}_V, \overline{S}_A, \overline{S}_H, \delta\right)$, where δ is the transmission time identifier.

(4) The above objective function can be jointly optimized,

$$h_V(\cdot), h_A(\cdot), h_H(\cdot) \sim \text{argmax}\, decode_V + decode_A + decode_H + \lambda \cdot Syn^2. \tag{7.13}$$

Finally, the optimized decoding models for each modality are used to decode the initial semantics of each modality $\overline{S}_V, \overline{S}_A, \overline{S}_H$.

Semantic compensation guided by inter-modal correlations

Due to the impact of loss, interference, and noise during the transmission, the quality of certain modality-specific initial semantic segments after decoding cannot be guaranteed. For example, in teleoperation tasks primarily based on haptic perception and feedback, the transmission of the residual semantics of visual and audio is relatively less important and is more prone to errors during transmission, resulting in relatively poor quality of the decoded initial visual and audio semantic segments. In tasks mainly involving audio and visual perception and feedback, the decoded haptic initial semantics also face a low-quality issue. In tasks where audio, video, and haptic perception and feedback are equally important, semantic segments of visual, audio, and haptic modalities may all have problems. Similar to the semantic synchronization process, there is still a lack of available prior information here. Given this, interrelationships among modalities' semantics should be further explored. Compensation for low-quality modality semantics can be achieved based on semantic information theory [19] and adversarial learning [21]. Specifically, taking the compensation of low-quality haptic semantic segments with visual semantic segments as an example, the process is provided in the following and shown in Fig. 7.7.

(1) The low-quality initial haptic semantic segment to be compensated, $\overline{\mathbf{S}}_H$, is input into the compensation model $r_H(\cdot)$, which can be composed of a deep neural network or a traditional machine learning model, to obtain the compensated haptic semantic segment $\overline{S}'_H$, *i.e.*, $\overline{S}'_H \sim r_H(\overline{S}_H)$.
(2) The normal-quality visual semantic segment $\overline{\mathbf{S}}_V$ is passed through a cross-modal generator $Gen(\cdot)$ to generate the corresponding haptic semantic segment $\overline{S}'_H$. That is, $\overline{S}'_H \sim Gen(\overline{S}_V)$. It should be noted that this generator can be obtained by fine-tuning a pre-trained deep neural network or can be constructed based on predefined cross-modal knowledge association rules.

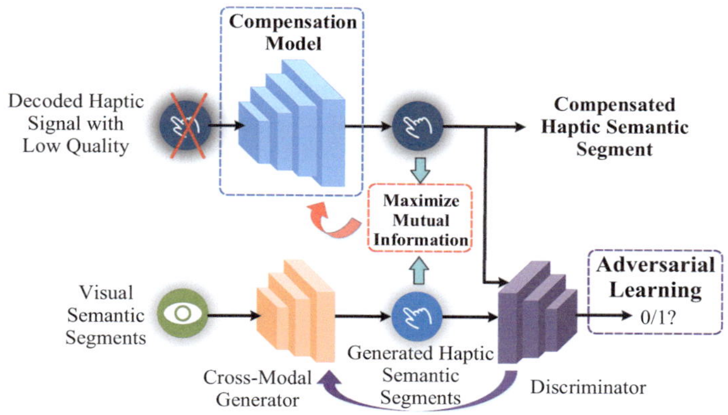

Fig. 7.7 Illustration of semantic compensation guided by inter-modal correlations

(3) Compared to signals, semantics are abstract concepts that cannot be measured using Euclidean distance or similar metrics. According to semantic information theory, the compensated haptic semantic segment $\overline{\mathbf{S}}'_H$ should be as close as possible to the haptic semantic segment $\overline{S}''_H$ generated from visual semantics. Based on this, the maximization of mutual information between the two is taken as a measure. That is, the optimization objective of the compensation model is designed as

$$\max I\left(\overline{S}'_H, \overline{S}''_H\right). \tag{7.14}$$

(4) $\overline{S}'_H$ and $\overline{S}''_H$ are passed through a discriminator based on the idea of adversarial learning. The role of the discriminator is to accurately determine whether the semantics of $\overline{S}'_H$ and $\overline{\mathbf{S}}'_H$ are close. By discriminating between the two, the cross-modal generator $Gen(\cdot)$ can be further fine-tuned and optimized.

(5) Based on the above idea, the optimization objective of the discriminator is designed as

$$\max Dis\left(\overline{S}''_H\right) - Dis\left(\overline{S}'_H\right), \tag{7.15}$$

where $Dis(\cdot)$ is the discrimination function. The optimization objective of the cross-modal generator is designed as

$$\max Dis\left(Gen\left(\overline{S}_V\right)\right). \tag{7.16}$$

By alternately iterating and optimizing the above objective functions, the training of the compensation model and the updating of the cross-modal generator are completed, thereby achieving haptic semantic compensation assisted by visual semantics. For low-quality initial audio semantic segments and low-quality initial visual semantic segments, the same mechanism can be applied to achieve compensation with the assistance of other modality semantics.

Destination source signal reconstruction based on semantic mapping

After completing semantic decoding for each modality and compensating for low-quality modality semantics, a corresponding knowledge base is formed based on the typical semantics and signals of each modality. Driven by techniques such as attention mechanism, a mapping from the semantic space of each modality to their respective signal space is realized, reconstructing the corresponding visual, audio, and haptic signals. This process is continuously optimized based on the evaluation of immersion in teleoperation experiences. Taking visual signal reconstruction based on visual semantic mapping as an example, the specific process is provided in the following and shown in Fig. 7.8.

(1) In the visual knowledge base construction phase, a set of typical visual semantics $\{Q_V\}$ and the corresponding real visual signal set $\{V\}$ are selected. Let Q_V be mapped through a visual semantic-to-signal mapping model $Rec(\cdot)$ to reconstruct visual signal $\widehat{V}$, that is, $\widehat{V} \sim Rec(Q_V)$.
(2) Based on semantic information theory, the comprehension deviation of V and $\widehat{V}$ for the visual semantics Q_V is obtained, that is,

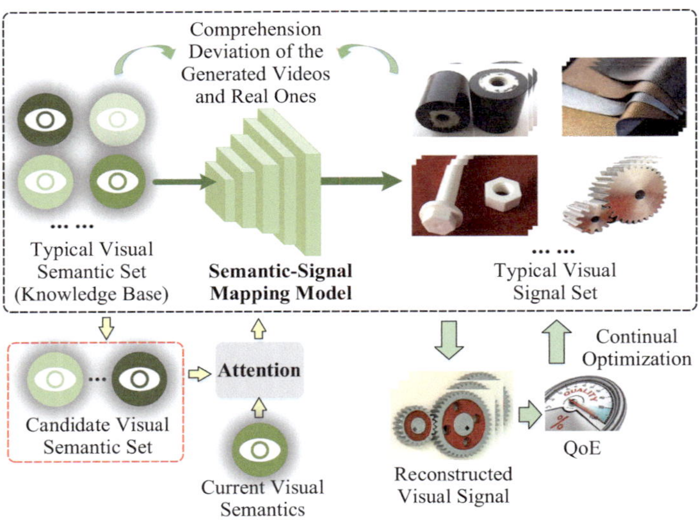

Fig. 7.8 Illustration of destination source signal reconstruction based on semantic mapping

$$\min I(V, Q_V) - I\left(\widehat{V}, Q_V\right). \tag{7.17}$$

Through optimization, the deviation is minimized to determine the structure and parameters of $Rec(\cdot)$, thereby completing the construction of the visual knowledge base at the master terminal.

(3) For the incoming visual semantics $\overline{S}'_V$, the K most similar typical visual semantics $\{Q_{V,1}, \ldots, Q_{V,K}\}$ are identified in the visual knowledge base. Based on the attention mechanism, the similarity $Sim(\overline{S}'_V, Q_{V,k})$ between $\overline{S}'_V$ and each typical visual semantics is calculated, and then normalized using $Norm(\cdot)$. Finally, the normalized coefficients are used to weight $\{Q_{V,1}, \ldots, Q_{V,K}\}$, which serves as the input $\overline{Q}_{V*}$ for $Rec(\cdot)$, that is,

$$\overline{Q}_{V*} = \sum_{k=1}^{K} Norm\left(Sim\left(\overline{S}'_V, Q_{V,k}\right)\right) \times Q_{V,k}. \tag{7.18}$$

(4) The trained visual semantic-to-signal mapping model outputs the reconstructed visual signal V^*, that is, $V^* \sim Rec(\overline{Q}_{V^*})$.

(5) For the audio and haptic modalities, the similar process is adopted to reconstruct the audio destination signal A^* from $\overline{S}'_A$, and the haptic destination signal H^* from $\overline{S}'_H$.

To further ensure the realism of the reconstructed destination signals on audio-visual and haptic feedback devices, a combination of subjective and objective evaluations is used for V^*, A^*, H^* to obtain the quality of experience (QoE) values for vision, audition, touch, and the overall teleoperation experience immersion. Specifically, for subjective evaluation, a certain number of operators can rate the experience; for objective evaluation, the reconstructed signals can be fed into a QoE model to predict their QoE values [22].

If the immersion in a particular modality is weak, for example, if the QoE value of the visual modality is low, the visual semantic-to-signal mapping model $Rec(\cdot)$ can be improved by constructing a more targeted typical visual semantics-signal set $\{Q_V, V\}$, refining the relationship between $\overline{S}'_V$ and $\{Q_{V,1}, \ldots, Q_{V,K}\}$, and continuously optimizing the visual destination signal reconstruction process based on visual semantic mapping. Through the above procedure, the rendering effect of the content and the immersive experience for the operator can be improved.

7.2 Cross-Modal Communications for Extremely Low Resource Scenarios

7.2.1 *Necessity and Challenges*

As described in the previous sections, existing cross-modal communication techniques are primarily designed for multi-modal service scenarios with relatively abundant communication and computing resources and a pursuit of immersive ultimate human experience, such as a holographic museum guide, rehabilitation training, and online immersive gaming. In the scenarios with extremely low communication and computing resources, it has significant bottlenecks. For example, when performing underwater equipment troubleshooting, challenges such as extremely low bandwidth, severe signal attenuation, and excessive energy consumption during the transmission of multi-modal signals exist [23]. In agricultural operations in remote mountainous areas, insufficient broadband network coverage and signal instability severely restrict the operational efficiency of robots [24]. In emergency rescue missions responding to natural disasters such as earthquakes, the teleoperation system must overcome complex issues, including a sudden reduction in computing power due to the destruction of communication infrastructure, and drastic fluctuations in the transmission environment caused by dramatic changes in terrain [25]. Importantly, Table 7.1 compares the differences between the existing regular cross-modal communications and the cross-modal communications for extremely low-resource scenarios. From this table, we can see, existing works about regular cross-modal communications may become unavailable in extremely low-resource scenarios.

It can be seen that existing cross-modal communication-related techniques cannot be directly applied to extremely low-resource scenarios. Taking the emergency rescue scenario as an example, the main difficulties faced are as follows:

Firstly, the effective bandwidth and device power consumption are severely limited, making existing multi-modal source signal compression and coding techniques inapplicable. In emergency rescue scenarios, the endurance capability of terminal robots is highly restricted, and the damage to communication base stations often results in a significant limitation of bandwidth. This poses a severe challenge for the compression and coding of multi-modal source signals that have been collected. Particularly for visual signals, it is necessary to ensure the integrity of key semantic details such as building cracks and structural deformations that are crucial for rescue decision-making, while also employing ultra-low-bit-rate transmission to reduce power consumption. On one hand, existing research typically compresses visual and haptic signals independently, failing to eliminate redundant information between the two modalities. On the other hand, current cross-modal compression methods are primarily designed for video and text, based on the assumption of global spatiotemporal-semantic alignment between them. However, in the context of emergency rescue, the haptic signals collected often correspond only to specific building components or areas in the videos. Therefore, constructing a strong correlation between haptic signals and local video areas is crucial. Additionally, achieving extreme compression

Table 7.1 Comparison of the regular cross-modal communications and the cross-modal communications for extremely low-resource scenarios

Aspect	Regular cross-modal communications	Cross-modal communications for extremely low-resource scenarios
Expected goal	Low-latency, high-reliability, and high-capacity multi-modal streaming, with a pursuit of immersive experience	Reliable and lightweight multi-modal streaming under extremely constrained communication and computing resources
Core element	Meeting the needs of haptic streaming, which has a greater impact on immersive experience; Haptic modality is assigned the highest priority in terms of transmission and processing	Audio and visual signals are more susceptible to the constraints of transmission resources and environmental factors; Haptic modality cannot take effect due to constrained audio-visual streaming
Supporting condition	There is a wealth of data and external knowledge available for reference; Communication infrastructure is well-developed with ample computational power in communication devices	In unknown and open environments, semantic label data and external knowledge are difficult to obtain; Communication infrastructure may be partially or completely damaged, and the computational power of communication devices may be limited
Realization method	Optimized cache management, communication resource allocation, and computation offloading algorithms are employed, as well as artificial intelligence-driven cross-modal signal processing methods	The collaborative transmission and processing method for multi-modal signals under extremely limited communication and computing resources almost remains a blank slate

of cross-modal source information while ensuring the integrity of key semantics becomes a major challenge faced by cross-modal communication research aimed at emergency rescue.

Secondly, the transmission environment is highly complex and dynamic. Existing transmission resource allocation strategies are inadequate for dealing with unknown conditions. In emergency rescue scenarios following disasters such as earthquakes, the communication environment is characterized by high dynamics and non-structured features. On one hand, the complex electromagnetic reflections caused by collapsed buildings exacerbate multi-path interference, while secondary disasters triggered by aftershocks can instantaneously alter channel attenuation characteristics, leading to abrupt degradation in the quality of communication links. On the other hand, the rescue site is often accompanied by power outages and overloaded emergency communication base stations, resulting in fragmented spectrum resources and extremely unstable energy supply. Efficient allocation of transmission resources is a crucial approach to ensuring transmission quality. However, both convex optimization-based and AI-based resource allocation strategies are predicated on the assumption that communication link state information is partially predictable or that channel changes are not drastic. These strategies are thus ill-equipped to handle the aforementioned extreme uncertainties. Therefore, designing transmission resource allocation strategies with strong adaptive adjustment capabilities for open and unknown environments is a crucial task. Moreover, rapidly and accurately matching the extremely valuable energy and spectrum resources with the heterogeneous transmission requirements of the audio, visual, and haptic streams to ensure transmission efficiency and stability is an extremely challenging issue.

Thirdly, the severe scarcity of external knowledge and computational power has led to a technological bottleneck in existing cross-modal signal recovery methods at the master domain. In emergency rescue scenarios, the haptic feedbacks received by the control devices and the quality of the videos ultimately presented are crucial for making timely and correct rescue decisions. However, existing cross-modal signal recovery methods are constrained by both a "knowledge desert" and a "computational power cage". On one hand, emergency rescue sites are often in an information-isolated state, unable to access external knowledge bases or collect sufficient usable data for the current environment. This lack of data for training severely affects the quality of signal recovery when using existing inter-modal correlation-based schemes. On the other hand, the control devices are often installed on mobile and portable platforms (such as mobile command vehicles or individual combat units), which have very limited computational power. In other words, implementing visual and haptic signal recovery faces a deeper semantic gap and higher computational complexity challenges. Therefore, how to recover high-quality detailed video and haptic contents from the extremely compressed bitstream received under the extremely scarce conditions of external knowledge and computational power of the control devices remains an unresolved issue.

7.2.2 *Scalable Cross-Modal Haptic Signal Recovery for Emergency Rescue*

Based on the above analysis, we take a specific study in [26] as an example to conduct a preliminary exploration in this direction. In an emergency rescue scenario, touch sensory and the related haptic signals are of great significance, as relying solely on audio-visual signals cannot provide operators with a comprehensive understanding of object properties, thereby affecting their fine manipulation. Currently, the quality of haptic signals in rescue robots is either low or even absent. The primary reason is that, compared with audio and visual signal acquisition, the main method for obtaining real haptic signals is more rigorous. It needs to use accelerometers or force sensors to capture the signals generated during the sliding of objects on surfaces, which reflect the textural characteristics of the objects. Due to the limitations of sensors, haptic signals are inherently difficult to obtain. More importantly, haptic streaming or the generation of virtual haptic sensations depends on the real-time and stable audio-visual streaming. However, in emergency environments, limited bandwidth and frequent bandwidth fluctuations restrict large-scale audio-visual streaming. Under such circumstances, the cross-modal haptic signal recovery or generation algorithm proposed in Chap. 5 fails. According to the actual needs of emergency rescue scenarios for haptic signal quality, a scalable cross-modal haptic signal recovery scheme (SCMHR) for extremely low-resource scenarios is proposed in this section [26]. The SCMHR can automatically adjust and select the optimal haptic signal recovery scheme according to the quality of the received audio-visual signals. Specifically, depending on different network conditions, the SCMHR will adaptively select the suitable scheme among coarse-grained haptic signal retrieval, fine-grained haptic signal retrieval, and haptic signal generation. In this way, it enables real time, lightweight haptic signal receiving and rendering, thereby fully leveraging the effectiveness and efficiency of the emergency rescue system.

Architecture

The architecture consists of three components: the device side with limited computing power, the wireless network, and the edge side situated adjacent to the robotic device. Each component will be described in detail below.

Device side: The device side typically comprises robots equipped with computational, sensing, and mobility capabilities, such as Unitree Go1, Unitree A1, and others. These robots are responsible for collecting and processing raw data, as well as executing commands from human operators. Subsequently, the fused features can be obtained by utilizing the correlation of multi-modal signals. Then, scalable encoding is employed to learn the structural semantic representation. Upon receiving the network state information sensed from the edge side, the device autonomously determines the relevant semantics for transmission. Additionally, the device performs quantization, channel coding, and modulation to facilitate multi-modal streaming over the wireless network.

Wireless network: The primary function of a wireless network is data transmission. On one hand, it promptly conveys the network state information sensed at the edge side to the device side. On the other hand, the device adaptively selects and transmits the optimal stream representation to the edge side via the wireless network.

Edge side: The edge side refers to the server located near the device side, providing caching, communication, and computing resources. Its responsibilities include assessing the network state information and promptly providing it to the device side, storing the existing haptic database, and generating haptic signals. The edge side receives semantic information at various granularities through the processes of demodulation, channel decoding, and inverse quantization. It then utilizes the semantics to recover haptic signals in three ways. (1) In scenarios with severely limited network bandwidth, the deep-layer feature maps of the fused features are transmitted to the edge side to determine coarse-grained category information. (2) As the bandwidth improves, the shallow-layer feature maps are transmitted and used to determine fine-grained category information. In both of the above cases, the haptic signals are recovered by retrieving signals of the same category from the existing database. (3) In scenarios with sufficient network bandwidth, the complete fused features are directly transmitted and used for haptic generation. The recovered haptic signals will be delivered to teleoperators via haptic feedback devices.

The SCMHR approach

Under this architecture, the SCMHR can be segmented into two main components according to their deployment locations. The first component, situated at the device side, contains networks E_V and E_A for extracting features from audio and visual signals, the feature fusion network F_{CMF} for obtaining semantics within audio and visual modalities. Since the functions of these two modules have been introduced in the preceding chapters, the specific details are omitted here. Moreover, it contains a scalable semantic encoding network E_S whose implementation detail will be highlighted. The second component, located at the edge side, comprises several networks: the scalable semantic decoding network D_S, the network state-awareness network D_{NS}, and the haptic recovery network D_H. D_S is paired with E_S. D_H further encompasses three main functions: the coarse-grained retrieval network D_{HCR}, the fine-grained retrieval network D_{HFR}, and the generation network D_{HG}. Since the functions of cross-modal retrieval and haptic signal generation have been respectively introduced in Sects. 5.1 and 5.4, the specific detail of D_H is also omitted here.

Scalable semantic encoding & decoding: After executing audio-visual feature extraction and feature fusion, the obtained fused feature is denoted as f_{V-A}. It amalgamates complementary information from both visual and audio signals. In order to adapt to different transmission environments, hierarchical semantic representations of f_{V-A} should be derived.

Specifically, this process is realized through the compact representation extraction for the fused features using neural networks. Feature maps at varying depths are employed to capture different levels of semantic granularity. The complete semantic representation of the fused features is denoted as f_{V-A}, while the shallow-layer and deep-layer feature maps extracted from the fused features via neural networks are represented as f_{V-A-s} and f_{V-A-d}, respectively. The feature f_{V-A} is designated for haptic generation, whereas f_{V-A-s} and f_{V-A-d} are used for fine-grained and coarse-grained retrieval, respectively. Coarse-grained retrieval encompasses a smaller number of broader categories, while fine-grained retrieval offers more specific and detailed classifications. For example, using f_{V-A} enables the generation of the "textile mesh" haptic signal. With f_{V-A-s} the object can be distinguished as "textile mesh" rather than "plastic mesh". f_{V-A-d} only allows classification within the broader category of "mesh", "metal", and "plastic". After acquiring the category information, haptic signals of the same category can be retrieved from the existing haptic database. Consequently, the semantic granularity progressively decreases from f_{V-A} to f_{V-A-s} and then to f_{V-A-d}, facilitating their natural compression into a scalable bitstream. Additionally, the bandwidth required to transmit these three types of semantic information is denoted as M_1, M_2, and M_3, respectively, with

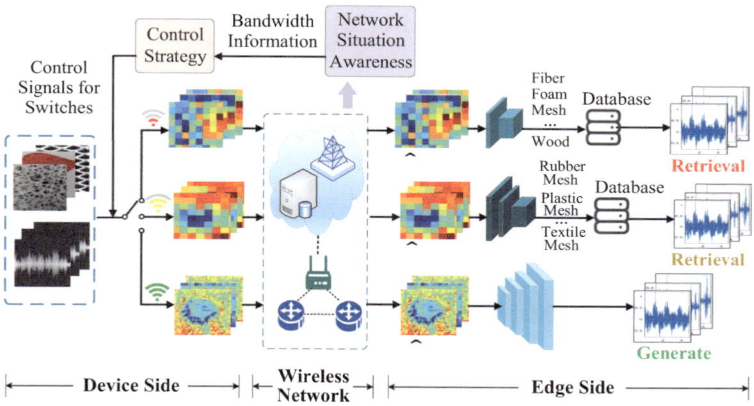

Fig. 7.9 Illustration of the scalable haptic recovery scheme based on network situation awareness

$M_1 > M_2 > M_3$. Finally, the process of the scalable semantic decoding is exactly the opposite of that of the scalable semantic encoding.

Network situation awareness: Network situational awareness refers to a comprehensive understanding of the current network environment, aimed at optimizing data transmission and enhancing service quality. The objective here is to efficiently transmit bit streams of varying scales to accommodate dynamic communication environments. Therefore, the primary focus is on network bandwidth, which can be obtained using the bang-on-bandwidth (BOB) method introduced in [27]. The core idea of the BOB is to utilize the collected information, including sending/receiving time and packet size, to assess both current and future bandwidth availability along the present path. Based on this, the device dynamically adjusts the scalable coding network according to the predicted bandwidth information. This process is visually depicted by adjusting the state of the switch, as shown in Fig. 7.9.

Based on the above description, the procedure of the SCMHR is provided in Algorithm 7.1.

Experimental result and analysis

The experiment simulates a rescue operation in an earthquake disaster area. Specifically, earthquakes often cause buildings to be damaged or collapse, trapping people inside. However, the instability and danger of the disaster zone prevent rescuers from directly entering to conduct rescue operations. Therefore, robots with multi-modal sensing and feedback capabilities can be deployed to operate in these hazardous environments. Here, various objects made of different materials are arranged in the scene to simulate potential rescue scenarios.

The constructed rescue prototype system is composed of two main components: the slave domain (device side) and the master domain (edge side), as shown in Fig. 7.10. The device side includes the Unitree A1 robot, which is designed to access

Algorithm 7.1 Procedure of the SCMHR [26]

Require: E_V, E_A, F_{CMF}, D_{NS}, E_S, D_{HCR}, D_{HFR}, D_{HG}, bandwidth required to transmit f_{V-A}, f_{V-A-s}, f_{V-A-d} (M_1, M_2, M_3)

The device side performs feature extraction and feature fusion using E_V, E_A, and F_{CMF}
The device side learns the hierarchical semantic representation through E_S
D_{NS} determines the network bandwidth and transmits it to the device side
if Bandwidth $> M_1$ **then**
  The device side transmits the fused features f_{V-A};
  The edge side performs the cross-modal haptic signal generation using D_{HG};
end if
if $M_2 <$ Bandwidth $< M_1$ **then**
  The device side transmits the shallow-layer feature map f_{V-A-s};
  The edge side performs the fine-grained cross-modal retrieval using D_{HFR} and searches the haptic signal in the existing database
end if
if $M_3 <$ Bandwidth $< M_2$ **then**
  The device side transmits the deep-layer feature map f_{V-A-d};
  The edge side performs the coarse-grained cross-modal retrieval using D_{HCR} and searches the haptic signal in the existing database
end if
The edge side delivers the recovered haptic signal to the teleoperators.

Output: The recovered haptic signal

hazardous areas that are inaccessible to humans. The robot is equipped with a depth camera and a microphone, enabling it to collect audio and visual signals from the environment [28]. The edge side mainly consists of a laptop equipped with an NVIDIA RTX 4080 and a haptic feedback device (Touch). The laptop uses the open-source Unitree Guide package to control the robot's movements. It is also responsible for presenting the received audio-visual signals and performing the tasks of decoding the fused semantics and generating haptic signals. The generated or recovered haptic signals are transmitted to the operator via the Touch device to enhance object perception and improve rescue efficiency. Communication between the master and slave terminals is conducted over a wireless network. The NetLimiter tool on the laptop adjusts the data reception rate to simulate various communication scenarios.

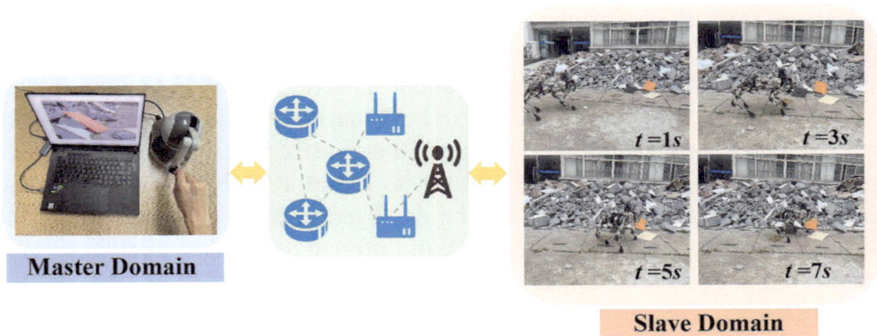

Fig. 7.10 Illustration of the constructed rescue prototype system

After constructing the prototype system, the scalability of the SCMHR is provided through haptic recovery experiments. These experiments use real network traces collected by the Federal Communications Commission, which are among the commonly used network traces [29]. These traces capture the variations in network bandwidth in real-world scenarios. Based on the characteristics of emergency rescue, 100 trace sets from this dataset are randomly selected, particularly those with bandwidths ranging from 2 to 40 Kbps. For bandwidths above 4.096 Kbps and below 8.192 Kbps, it performs coarse-grained cross-modal retrieval. For bandwidths between 8.192 and 16.384 Kbps, it switches to fine-grained cross-modal retrieval. For bandwidths exceeding 16.384 Kbps, it directly generates haptic signals. It is important to note that these boundaries are not rigid and are influenced by the dimensions of features, the distribution of features, and the corresponding channel coding strategy. In Fig. 7.11, six representative traces are selected and their bandwidth fluctuations are recorded. For each bandwidth condition, four representative moments are chosen, and the haptic signals recovered by the SCMHR are demonstrated. The experimental results indicate that the SCMHR can reliably recover haptic signals under varying network conditions by employing different strategies, highlighting its scalability.

7.3 Summary

In this chapter, we introduce two extended techniques for cross-modal communications. Firstly, we present a novel communication paradigm named cross-modal semantic communications, focusing on semantics rather than signals. Compared with semantic communication, cross-modal semantic communication not only concerns semantics within each modality but also fully considers the semantic correlation between visual, audio, and haptic modalities. That is, the term "cross-modal" guides the perception, transmission, and reception of semantics. While eliminating information redundancy, it ensures transmission efficiency and reliability, providing users with an immersive experience. Specifically, taking the immersive teleoperation as an example, we systematically provide the implementation methods for cross-modal semantic extraction at the slave domain, cross-modal semantic transmission between the master and slave domains, and cross-modal semantic recovery at the master domain. However, this is just one way of its realization. The research on cross-modal semantic communications is still rapidly developing. Secondly, we describe details about cross-modal communication under extremely low-resource scenarios. Compared with regular cross-modal communications, it has significant differences in the aspects of expected goal, core element, supporting condition, and realization method, and there are also many challenges to be solved. Based on this, we take the haptic signal recovery in the emergency rescue scenario as an example to illustrate a feasible solution. Specifically, an architecture for edge-device collaboration and a scalable cross-modal haptic recovery scheme are proposed. This scheme implements the optimal recovery way based on the received audio-visual modalities of different granularities, ensuring real-time acquisition of coarse-grained or fine-grained haptic

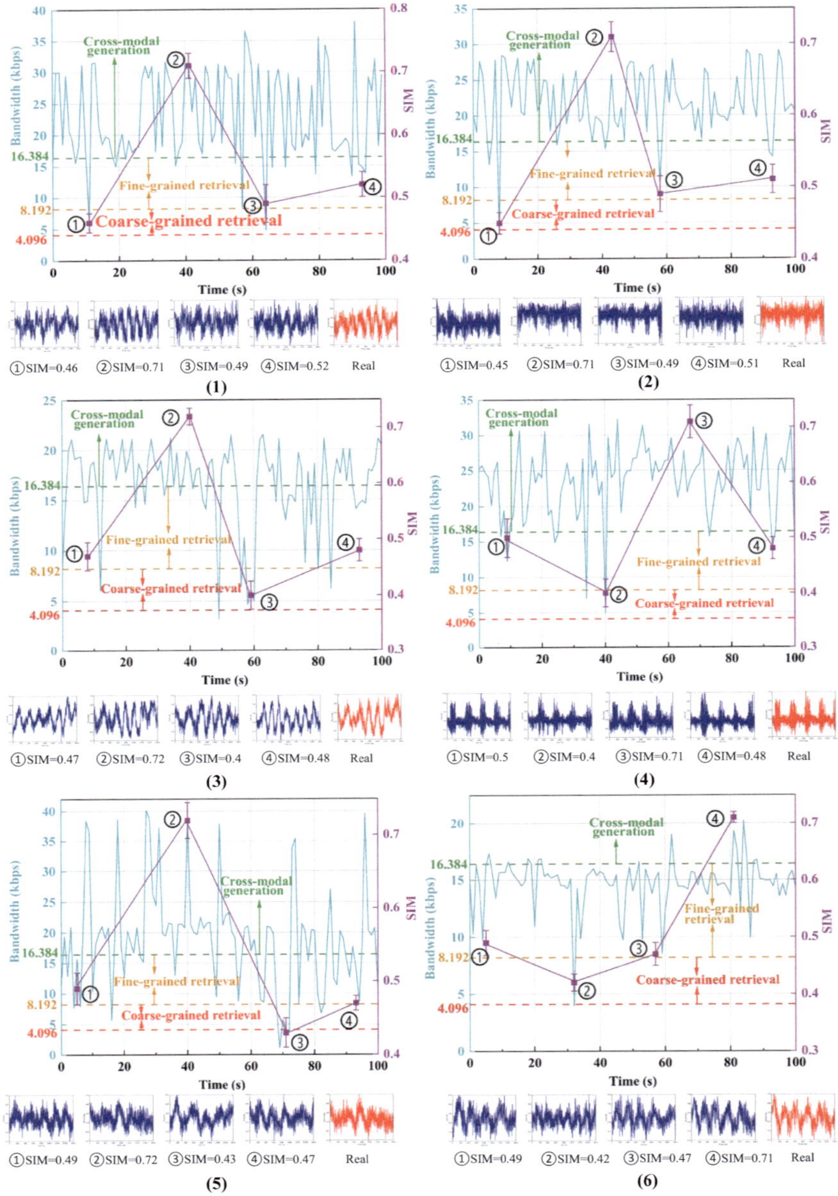

Fig. 7.11 The haptic signal recovery performance of the SCMHR under six different bandwidth conditions

feedback. Through the construction of a prototype system and experiments, the proposed scheme can reliably recover haptic signals in an emergency rescue scenario with extremely low transmission and computing resources.

References

1. Getu TM, Kaddoum G, Bennis M (2024) Semantic communication: a survey on research landscape, challenges, and future directions. Proc IEEE 112(11):1649–1685
2. Yang W, Du H, Lie Z, Lim WYB, Xiong Z, Niyato D (2023) Semantic communications for future internet: fundamentals, applications, and challenges. IEEE Commum Surv Tut 25(1):213–250
3. Liao J, Wei X, Zhou L (2019) Artificial intelligence-driven cross-modal semantic communication system. ZTE Technol J 30(S1):33–39
4. Li A, Wei X, Wu D, Zhou L (2022) Cross-modal semantic communications. IEEE Wirel Commun 29(6):144–151
5. Zhou L, Wu D, Chen J, Wei X (2020) Cross-modal collaborative communications. IEEE Wirel Commun 27(2):112–117
6. Wei X, Zhou L (2021) AI-enabled cross-modal communications. IEEE Wirel Commun 28(4):182–189
7. Erdemir E, Tung TY, Dragotti PL, Gündüz D (2023) Generative joint source-channel coding for semantic image transmission. IEEE J Sel Area Comm 41(8):2645–2657
8. Zhang G, Hu Q, Qin Z, Cai T, Yu G, Tao X (2024) A unified multi-task semantic communication system for multimodal data. IEEE Trans Commun 72(7):4101–4116
9. Zhao L, Wu D, Zhou L (2023) Data utilization versus privacy protection in semantic communications. IEEE Wirel Commun 30(3):44–50
10. Jankowski M, Gündüz D, Mikolajczyk K (2021) Wireless image retrieval at the edge. IEEE J Sel Area Comm 39(1):89–100
11. Xie H, Qin Z, Tao X, Letaief KB (2022) Task-oriented multi-user semantic communications. IEEE J Sel Area Comm 40(9):2584–2597
12. Guo J, Chen H, Song B, Chi Y, Yuen C, Yu FR, Li GY, Niyato D (2024) Distributed task-oriented communication networks with multimodal semantic relay and edge intelligence. IEEE Comm Mag 62(6):82–89
13. Koller D, Friedman N (2009) Probabilistic graphical models: principles and techniques. The MIT Press
14. Purri M, Dana K (2020) Teaching cameras to feel: Estimating tactile physical properties of surfaces from images. In: Proceedings of the European conference on computer vision, online, Aug 23–28, pp 1–20
15. Hou Z, She C, Li Y, Niyato D, Dohler M, Vucetic B (2021) Intelligent communications for tactile internet in 6G: requirements, technologies, and challenges. IEEE Comm Mag 59(12):82–88
16. Zou J, Li C, Liu C, Yang Q, Xiong H, Steinbach E (2020) Probabilistic tile visibility-based server-side rate adaptation for adaptive 360-degree video streaming. IEEE J-STSP 14(1):161–176
17. Wei X, Zhao J, Zhou L, Qian Y (2020) Broad reinforcement learning for supporting fast autonomous IoT. IEEE Internet Things J 7(8):7010–7020
18. Chen CLP, Liu Z (2018) Broad learning system: an effective and efficient incremental learning system without the need for deep architecture. IEEE Trans Neural Netw Learn Syst 29(1):10–24
19. Zhong Y (2017) A theory of semantic information. China Commun 14(1):1–17
20. Liu X, Zhang F, Hou Z, Mian L, Wang Z, Zhang J, Tang J (2023) Self-supervised learning: generative or contrastive. IEEE Trans Knowl Data Eng 34(1):857–876
21. Goodfellow I, Pouget-Abadie J, Mirza M, Xu B, Warde-Farley D, Ozair S, Courville A, Bengio Y (2023) Generative adversarial networks. Commun ACM 63(11):139–144

22. Wei X, Zhou L (2019) Multimedia QoE evaluation. Springer, Berlin
23. Sun J, Zhang Q, Lu Y, Huang B, Li Q (2025) A review of touching-based underwater robotic perception and manipulation. Machines 13(1):41
24. Chen J, Ma W, Liao H, Lu J, Yang Y, Qian J, Xu L (2024) Balancing accuracy and efficiency: the status and challenges of agricultural multi-arm harvesting robot research. Agronomy-Basel 14(10):2209
25. Yao Z, Cheng W, Zhang W, Zhang H (2021) Resource allocation for 5G-UAV-based emergency wireless communications. IEEE J Sel Area Comm 39(11):3395–3410
26. Liu H, Wei X, Gao Y, Zhou L, Qian Y (2025) Cross-modal haptic generation for emergency rescue in internet of robotic things. IEEE Internet Things 12(12):21415–21428
27. Bentaleb A, Akcay MN, Lim M, Begen AC, Zimmermann R (2023) BoB: bandwidth prediction for real-time communications using heuristic and reinforcement learning. IEEE Trans Multimedi 25:6930–6945
28. Sombolestan M, Nguyen Q (2024) Adaptive-force-based control of dynamic legged locomotion over uneven terrain. IEEE Trans Robot 40:2462–2477
29. Burger EW, Krishnaswamy P, Schulzrinne H (2023) Measuring broadband America: a retrospective on origins, achievements, and challenges. SIGCOMM Comput Commun Rev 53(2):11–21

Chapter 8
Conclusion

8.1 Concluding Remarks

To support the thriving multi-modal services, the cross-modal communication paradigm is proposed by the MultiMedia Communication (MMC) Research Group, which comprises researchers from Nanjing University of Posts and Telecommunications and the Army Engineering University of PLA. This book offers a systematic and in-depth investigation into this highly promising research area. There are several specific concluding remarks given in the following.

Firstly, this book provides a deep introduction to the characteristics of the current and future multi-modal services. Based on these characteristics, a general cross-modal communication architecture is constructed. Three key techniques, their encountered challenges, and design principles are also analyzed. It can be concluded that both the architecture construction and the key techniques design should center on the core notion of potential correlations among modalities or "semantic". In other words, it should focus on uncovering, analyzing, and utilizing semantic information that reflects the potential correlations among modalities.

Secondly, this book provides a detailed description of the three key techniques in cross-modal communications: cross-modal coding, cross-modal transmission, and cross-modal signal recovery. For *cross-modal coding*, the focus is on eliminating content redundancy across modalities and on coding theory and codec design. It elaborates on the technical details and experimental results of the proposed methods. Notably, it derives the minimum bits required for haptic signal compression under visual stream rates and the optimal coding rate for visual-haptic compression. Based on these findings, corresponding codecs are designed. By leveraging semantic information between modalities, it achieves significant performance gains over existing separate coding schemes. For *cross-modal transmission*, it takes into account the importance of content to be transmitted, the characteristics and requirements of heterogeneous streams, and the dynamic transmission environment with

X. Wei et al., *Cross-Modal Communication Technology*, Wireless Networks,
https://doi.org/10.1007/978-3-032-00957-9_8

limited resources. Existing transmission schemes are categorized into preemptive-resume and non-preemptive-resume scheduling. The former uses correlations within or between modalities to handle competitions of multi-modal streams caused by priority differences. The latter achieves efficient multi-modal streaming through tailored scheduling schemes and switching mechanisms, enhanced by AI methods and a cloud-edge-terminal collaborative framework. For *cross-modal signal recovery*, it presents different recovery schemes based on the specific multi-modal service types, such as audio-visual-aided haptic signal recovery, haptic-aided visual signal recovery, and visual-haptic mutual signal recovery. These schemes focus on exploring potential correlations between modalities to enable semantic information transfer and mapping between intact and damaged modalities or segments. Additionally, depending on the data distribution and requirements of specific multi-modal services, signal recovery can also be realized by methods such as information retrieval or cross-modal super-resolution signal reconstruction, which are taken as effective supplements to classic recovery schemes.

Thirdly, on the basis of theoretical derivation and key technology design, this book provides a preliminary exploration of the aspect of practical application. It introduces a large-scale multi-modal database developed by the MMC Research Group, which can be used for theoretical research and algorithm verification in the cross-modal communication area. Furthermore, it presents three developed prototype systems: a visual-haptic human-machine interactive system, a virtual acupuncture skill training application, and a remote throat swab sampling platform. These prototype systems demonstrate significant potential in the domains of remote industrial control, education, and healthcare, and provide valuable insights for developing other multi-modal service-related systems.

Fourthly, based on the regular cross-modal communication technology, this book presents two extended directions in this area. On one hand, it describes cross-modal semantic communications, combining cross-modal communications with semantic communications. The cross-modal semantic communication concerns semantics rather than signals. It also leverages both intra-modal and inter-modal semantics to address the ambiguity and polysemy issues in existing semantic communications, thereby enhancing transmission reliability. On the other hand, it points out that for extremely low-resource scenarios like underwater equipment troubleshooting, agricultural operations in remote mountainous areas, emergency rescue missions responding to natural disasters, regular cross-modal communications often fail, and the related key techniques need to be re-designed. This book takes cross-modal haptic signal recovery in emergency rescue missions as an example, presenting viable solutions, including an edge-device collaboration architecture and a scalable recovery scheme.

8.2 Future Work

Cross-modal communication technology is a promising research topic now, which is greatly significant for the emerging multi-modal services in the B5G and 6G era. This book provides guidance on several important aspects of cross-modal communications. There are still many studies that need to be carefully considered in the future.

At present, multi-modal services mainly focus on auditory, visual, and touch sensories and the related signals. However, there exist the other two important sensories: olfactory and gustatory. Once olfaction and gustation are integrated into multi-modal services, existing cross-modal coding, transmission, and signal recovery schemes may need to be re-designed. Moreover, previous studies have mostly relied on signal processing and AI methods to explore the semantics among modalities. Nevertheless, from a neuroscience perspective, there are natural and intrinsic interactions among the human sensory subsystems of auditory, visual, touch, olfactory, and gustatory modalities, which jointly shape human's experience. If these interactions can be fully utilized, they will highly promote theoretical and applied researches in cross-modal communications.

Although the essential nature of "semantics" in cross-modal communications has been studied, it is still far from sufficient. At the fundamental theoretical level, there is a need for a quantitative description of semantics and its impact on the performance of the designed cross-modal communication-related techniques. For example, it is necessary to determine the limits of multi-modal streaming rates under any semantic distortion, as well as the limits of semantic deviation under any multi-modal streaming rate. At the key technical research level, there has been no specific coding scheme that combines intra-modal and inter-modal semantics. In addition, how to utilize the potential correlations among modalities to achieve cross-modal transmission and signal recovery in scenarios with extremely low resources and scarce external data/prior knowledge remains challenging. This requires further exploration and practice in the future.

The core idea of cross-modal communication technology has the potential to provide inspiration for solving key issues in other fields. For instance, in the field of smart education, the online-merge-offline learning has now become mainstream. Under these circumstances, knowledge delivery can be achieved through both online learning platforms and offline classroom teachers, while students can generate feedback, such as learning behaviors and testing results. Due to the significant differences in the characteristics of online and offline learning scenarios, they can be regarded as two different types of modality streams. Previous studies have mostly conducted these two types of learning activities separately, leading to obstacles in information exchange. If the core idea of cross-modal communications can be utilized, it is expected to break down these obstacles, achieve deep integration of online learning and offline teaching, and thereby enhance students' learning outcomes and experiences.

MIX
Papier aus verantwortungsvollen Quellen
Paper from responsible sources
FSC® C105338

If you have any concerns about our products,
you can contact us on
ProductSafety@springernature.com

In case Publisher is established outside the EU,
the EU authorized representative is:
Springer Nature Customer Service Center GmbH
Europaplatz 3, 69115 Heidelberg, Germany

Printed by Libri Plureos GmbH
in Hamburg, Germany